SpringerBriefs in Materials

Series Editors

Sujata K. Bhatia, University of Delaware, Newark, DE, USA

Alain Diebold, Schenectady, NY, USA

Juejun Hu, Department of Materials Science and Engineering, Massachusetts Institute of Technology, Cambridge, MA, USA

Kannan M. Krishnan, University of Washington, Seattle, WA, USA

Dario Narducci, Department of Materials Science, University of Milano Bicocca, Milano, Italy

Suprakas Sinha Ray , Centre for Nanostructures Materials, Council for Scientific and Industrial Research, Brummeria, Pretoria, South Africa

Gerhard Wilde, Altenberge, Nordrhein-Westfalen, Germany

The SpringerBriefs Series in Materials presents highly relevant, concise monographs on a wide range of topics covering fundamental advances and new applications in the field. Areas of interest include topical information on innovative, structural and functional materials and composites as well as fundamental principles, physical properties, materials theory and design.

Indexed in Scopus (2022).

SpringerBriefs present succinct summaries of cutting-edge research and practical applications across a wide spectrum of fields. Featuring compact volumes of 50 to 125 pages, the series covers a range of content from professional to academic. Typical topics might include

- A timely report of state-of-the art analytical techniques
- A bridge between new research results, as published in journal articles, and a contextual literature review
- A snapshot of a hot or emerging topic
- An in-depth case study or clinical example
- A presentation of core concepts that students must understand in order to make independent contributions

Briefs are characterized by fast, global electronic dissemination, standard publishing contracts, standardized manuscript preparation and formatting guidelines, and expedited production schedules.

N. Sukumar

Navigating Molecular Networks

 Springer

N. Sukumar 🆔
School of Artificial Intelligence
Amrita Vishwa Vidyapeetham
Coimbatore, Tamil Nadu, India

ISSN 2192-1091   ISSN 2192-1105 (electronic)
SpringerBriefs in Materials
ISBN 978-3-031-76289-5   ISBN 978-3-031-76290-1 (eBook)
https://doi.org/10.1007/978-3-031-76290-1

This Springer imprint is published by the registered company Springer Nature Switzerland AG
The registered company address is: Gewerbestrasse 11, 6330 Cham, Switzerland

If disposing of this product, please recycle the paper.

*Dedicated to the memory of my wife Sunanda,
my mother Smt. Rajeshwari, my uncle
Shri E.G. Vaidyanathan, and my aunts
Dr. Annapurna and Smt. Sivakamu
Vaidyanathan.*

Foreword

Chemical space is a conceptual space inhabited all possible compounds known and unknown. It is where we can find solutions to all our pressing problems in technology, environment and healthcare. Unfortunately the space is as vast as the physical sea of outer space and navigating such a space to find islands of hope is humanly impossible. It is where Artificial Intelligence based algorithmic probes and bots help us. The Question then is: Should AI algorithms / bots learn on their own, or do they need to be taught the physical laws of the universe to search effectively? And how can human chemists and AI work together effectively?

The new generation chemists and chemical engineers need to be exposed to those issues along with the existing mathematical frameworks and tools that modern AI employs for searching such uncharted spaces. Professor Sukumar, who is a pioneer in this field with decades of experience, is the right person to give guidance to youngsters venturing into this adventurous intellectual sport. I have the opportunity to sit in his class and educate myself on how the ML and Generative AI is applied to cheminformatics. So I am sure that the students will enjoy reading and using this book. It has all the ingredients that ignite a flame of curiosity and excitement.

I wholeheartedly recommend this book to students, teachers and researchers. For teachers this book constitutes an excellent didactic instrument.

Prof. Soman K. P.
Dean, School of Artificial Intelligence
Amrita Vishwa Vidyapeetham
Coimbatore, India

Preface

Molecules can be related to one another in multiple ways—through similarity relationships, through physical interactions or through chemical transformations, giving rise to different kinds of molecular networks, with different characteristics. This book explores the theoretical principles underlying the treatment of molecular networks and "chemical space", the abstract space of all possible physically realizable molecules, from vector space, graph theoretic and data science perspectives. While there are many texts and compilations on various aspects and selected topics from among those covered in this book, such as graph theory, signaling and regulatory networks, random matrix theory, and deep learning, and while there are tantalizing threads of connections running through the primary literature, there seems to be no other book or resource that treats all this related material in a unified, coherent and logically connected manner. Key features include: (1) presentation from vector space, graph theoretic and data science perspectives; (2) exploration of different kinds of molecular networks; (3) discussion of Generative Deep Learning models. The subject matter straddles the disciplines of physics, chemistry, materials science and data science/AI. The emphasis in the book is on conceptual treatment, rather than on surveying specific applications.

While the Hohenberg-Kohn theorem guarantees a one-to-one mapping between the external potential (represented by the atomic coordinates in Coulombic systems and materials) and the ground state electron density and material properties, the inverse problem—that of discovering the molecular structure corresponding to a specific range of material properties—has been elusive and was long thought to be intractable. But in recent years, the use of artificial intelligence (AI) and machine learning (ML) techniques to leverage the power of first principles (DFT) calculations, and the advent of generative AI models to explore the compositional space of new materials promises to overturn this conventional wisdom. This book explores the theoretical principles underlying the modern graph theoretic, AI and ML techniques being increasingly used in the design of new materials with specific properties.

After a brief introduction to molecular networks and different kinds of graphs/ networks in Chap. 1, we will delve into the mathematical fundamentals required for developing the concepts in the book, first from a vector space viewpoint in Chap. 2, then from the network or graph theoretical perspective in Chaps. 3 and 4, and putting the two together in the following two chapters. Chapter 5 deals with visualizing and navigating chemical space networks. Chapter 6 delves into generative AI and growing the network, in the process bringing out the threads of connections between diverse topics such as similarity kernels, Gaussian processes, spectral graph theory, random matrix theory and deep learning. We close with some philosophical thoughts on discovery and creativity in Chap. 7.

Huran, India   N. Sukumar
September 2024

Acknowledgments

First of all I thank Prof. B. Ananthanarayan for inspiring and encouraging me to write this book.

I am deeply indebted to the late Prof. Curt Breneman, the late Prof. Santosh Kumar, Prof. Sudeepto Bhattacharya, Prof. V. K. Jayaraman, Prof. K. P. Soman, Prof. Sai Sundarakrishna, Dr. Vinith R, Dr. Pratiti Bhadra, Dr. Ganesh Prabhu, Dr. Michael Krein, Dr. Rick Rejeleene, Manuja Kothiyal and S. Sowmya for many helpful discussions without which this book would not have been possible.

I am also grateful to Prof. Areejit Samal and Dr. Vivek Ananth RP for helpful discussions and for sharing data prior to publication.

About This Book

The book is addressed at senior undergraduate and graduate students in physics, chemistry, and materials science, as well as anyone interested in learning about how computational, network, AI and ML techniques are poised to transform the search for new materials. This book explores the theoretical principles underlying the treatment of molecular networks and "chemical space", the abstract space of all possible physically realizable molecules, from vector space, graph theoretic and data science perspectives, bringing out the tantalizing threads of connections between diverse topics such as Similarity Kernels, Gaussian Processes, Network measures, Spectral Graph Theory and Random Matrix theory. Visualization of molecular networks and navigating these networks also receive due attention. This is followed up with an exploration of modern Generative Deep Learning models that are being increasingly used in the search for new materials with specific properties, and some of the most exciting developments in this area. The book closes with a discussion on the meanings of discovery and creativity, and the role of artificial intelligence (AI) therein.

Contents

About the Author

N. Sukumar is Adjunct Professor at the School of Artificial Intelligence, Amrita Vishwa Vidyapeetham, Coimbatore, India. Earlier, he was Professor and Founding Head of the Department of Chemistry and Founding Director of the Center for Informatics at Shiv Nadar University, India. He got his M. Sc. in Chemistry from the Indian Institute of Technology, Kanpur, Ph.D. from the State University of New York at Stony Brook and postdoctoral appointments at the University of Southern California, the University of New Orleans and Marquette University. He was Alexander von Humboldt Fellow at the University of Bonn, in Germany, and served as visiting scientist at the Wadsworth Institute of the New York State Department of Health, and as Associate Research Professor at the Rensselaer Polytechnic Institute in Troy, NY. His research interests include quantum chemistry, density functional theory, development of novel molecular descriptors and robust computational and cheminformatics methods for the discovery of molecules and materials with specific chemical and biological properties. Currently active research programs include: drug, polymer and nanomaterials design through informatics and deep learning; chemical and biological networks—developing novel ways of looking at chemical space, using graph and network properties for the study and design of molecular libraries; and materials design using machine learning and first-principles computations, to discover complex quantitative relationships between structure and properties of materials. Sukumar is also the editor of *A Matter of Density: Exploring the Electron Density*

Concept in the Chemical, Biological, and Materials Sciences, published by John Wiley & Sons in 2012, and has co-authored *Computational Drug Discovery—A Primer*, published by Ion Cures Press in 2023, besides authoring several other book chapters in various books and research papers, and two books on photography.

Chapter 1
Molecular Networks

To see a world in a grain of sand...
—William Blake, Auguries of Innocence [1]

1.1 Why Molecular Networks? Graphs and Simplices

Humanity has always been fascinated by the structure and constitution of diverse materials; we have admired and collected gems, precious stones, brightly colored minerals, and crystals for millennia. However, the scientist seeks organizing principles behind the wealth of information collected by our senses (and increasingly nowadays, by instrumentation). We represent these organizing principles as relationships between different objects or materials. These relationships can be of similarity (such as crystals with similar shapes or gemstones with similar color), of transformation (where one material transforms into another upon heating or reacting with water), or of other kinds. Such relationships can be conveniently depicted (both visually and mathematically) as edges of a graph. A *graph* is a collection of objects (represented by *nodes* or *vertices*) and pairwise relationships between them (represented by *edges* connecting the nodes in pairs). Mathematically [2], a graph is an ordered triple representation of an algebraic object consisting of a non-empty set (U, E, Ψ_G), with nodes $U = \{u_i | i = 1, 2, 3, \ldots, n\}$, edges $E = \{e_i | i = 1, 2, 3, \ldots, m\}$, such that $U \cap E = \emptyset$, and an incident function defined by $\Psi_G : E \to [U]^2; e \mapsto \Psi_G(e) = \{u_i, u_j\}$. A network is a four-tuple representation of a dynamical object $G = (U_t, E_t, \Psi_{N_t}, J_t)$, where t is a (real or simulated) time parameter, and J_t is the activity of edges and vertices of the network with time. Thus a network also has a time dimension to it, but following common parlance, we will often use the terms graph and network synonymously whenever such usage is unlikely to give rise to confusion.

Graph theory has a long and rich history in chemistry, even pre-dating modern structural ideas and the concept of molecules [3]. In modern usage, a molecule or crystal is commonly visualized as a set of atoms (forming the nodes of the molecular graph) connected by bonds (which constitute the edges of the graph), as in Fig. 1.1a. But we can also visualize graphs or networks whose nodes are individual molecules,

© The Author(s), under exclusive license to Springer Nature Switzerland AG 2024
N. Sukumar, *Navigating Molecular Networks*,
SpringerBriefs in Materials, https://doi.org/10.1007/978-3-031-76290-1_1

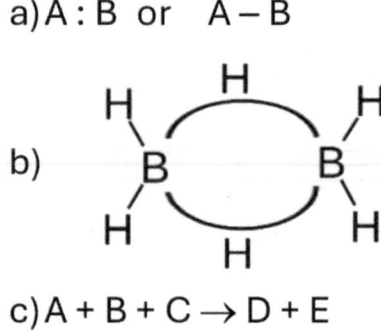

a) A : B   or   A − B

b)

c) A + B + C → D + E

Fig. 1.1 a) The Lewis electron-pair bond is formed by a pair of electrons with opposite spin (shown here as the pair of dots) being shared between a pair of atoms A and B. **b)** The three-center bonds or banana bonds (shown here as curved lines) in diborane are each formed by a pair of electrons with opposite spin shared by three atoms: two boron atoms and one hydrogen. The four peripheral B–H bonds are covalent two-center two-electron bonds. **c)** A chemical reaction transforms reactants (here A, B and C) into products (here D and E)

molecular fragments, or materials, and whose edges represent some pairwise relationship between the nodes. The relationship can be conceptual (such as a similarity relationship), physical (such as an interaction or physical contact), or functional (such as give rise to regulatory networks). Much of the discussion in this book will deal with such chemical networks.

Of course, one can also envisage relationships that are not just pairwise—the three-center ("banana") bonds [4, 5] in boranes (shown in Fig. 1.1b) being a familiar example; other multi-center bonds or many-body interactions are also now well-known. Another example is a reaction network, where a chemical reaction transforms a collection of molecules (reactants) into a different collection of molecules (products); see Fig. 1.1c.

Such many-body relationships require a generalization of graph theory to simplices or hypergraphs. A triangle that connects three vertices is a 2-simplex, a simple shape requiring two dimensions for its depiction; a tetrahedron that connects four vertices is a 3-simplex, requiring three dimensions for its depiction. We shall not deal with these objects further in most of this book, but we will return to them briefly in Sects. 4.2 and 6.8.

1.2   Matrix Representations of Weighted and Unweighted Networks

Any graph (with pair-wise connections) consisting of n nodes can also be represented as an n × n *adjacency matrix* **A**, where each row and each column represents a different node. In a *simple graph*, i.e. one without self-loops (edges connecting a node to itself), the diagonal elements are all zero:

$$A_{ii} = 0, \tag{1.1}$$

while any off-diagonal element $A_{ij} = 1$ if the corresponding nodes i and j share an edge, and 0 otherwise, as shown in Fig. 1.2. The corresponding adjacency matrix, shown in Table 1.1a, represents an unweighted graph, with all matrix elements being either zero or one. In an unweighted graph, all edges are similar, and the matrix elements of the adjacency matrix are restricted to the values 0 (signifying no edge between the corresponding pair of nodes) or 1 (signifying the presence of an edge between the corresponding pair of nodes). The adjacency matrix is symmetric about the main diagonal:

$$A_{ij} = A_{ji}. \tag{1.2}$$

The number of connections a given node has with other nodes in the graph is known as the *degree* of the node:

$$k_i = \sum_{j=1}^{n} A_{ij}. \tag{1.3}$$

Molecular graphs can also be represented as SMILES [6, 7] (Simplified Molecular Input Line Entry System) strings, a popular string notation employing the ASCII charter set, that is easily readable by humans, as well as being machine readable. For instance, the molecule shown in Fig. 1.2 can be represented as the SMILES string: NCC(=O)C. SMILES strings can represent straight-chain, branched, as well as ring systems, and can be converted into a molecular graph or adjacency matrix. They can distinguish aliphatic from aromatic rings, and distinguish between cis–trans stereoisomers as well enantiomers. Although a SMILES string is not unique for a given molecule, it can be readily converted into a Canonical SMILES string that is unique. As we shall see in Sect. 6.4 and 6.5, the SMILES molecular grammar can be easily learnt by a language model.

In a *weighted graph*, each edge between nodes i and j is associated with a weight w_{ij}, and the matrix elements of the weighted adjacency matrix can take on any real number value associated with the weight w_{ij}. An example of this is the atomic

Fig. 1.2 Molecular structure represented by the matrices in Table 1.1, with the atoms numbered. Unlabeled atoms are carbons

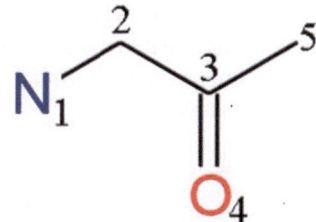

Table 1.1 (a)Adjacency matrix **A** for the molecule shown in Fig. 1.2, (b) atomic number bond order matrix, (c)Laplacian matrix $\mathbf{L} = \mathbf{D} - \mathbf{A}$, (d)Molecular incidence matrix **M**, (e) the matrix **Q** $= \mathbf{MM}^T = \mathbf{D} + \mathbf{A}$, and (f) Oriented incidence matrix **N** for the same molecule

(a) Adjacency Matrix

A	1	2	3	4	5
1	0	1	0	0	0
2	1	0	1	0	0
3	0	1	0	1	1
4	0	0	1	0	0
5	0	0	1	0	0

(b) Atomic Number-Bond Order Matrix

AtNum.BO	1	2	3	4	5
1	7	1	0	0	0
2	1	6	1	0	0
3	0	1	6	2	1
4	0	0	2	8	0
5	0	0	1	0	6

(c) Laplacian Matrix

L	1	2	3	4	5	Row Sum
1	1	-1	0	0	0	0
2	-1	2	-1	0	0	0
3	0	-1	3	-1	-1	0
4	0	0	-1	1	0	0
5	0	0	-1	0	1	0

(d) Incidence Matrix

Atoms

M	Edges 1	2	3	4	Row Sum
1	1	0	0	0	1
2	1	1	0	0	2
3	0	1	1	1	3
4	0	0	1	0	1
5	0	0	0	1	1

(e) $\mathbf{Q} = \mathbf{MM}^T = \mathbf{D} + \mathbf{A}$

L	1	2	3	4	5
1	1	-1	0	0	0
2	-1	2	-1	0	0
3	0	-1	3	-1	-1
4	0	0	-1	1	0
5	0	0	-1	0	1

(f) Oriented Incidence Matrix

Atoms

M	Edges 1	2	3	4
1	1	0	0	0
2	-1	1	0	0
3	0	-1	1	1
4	0	0	-1	0
5	0	0	0	-1

number-bond order matrix, shown in Table 1.1b for the same molecule (Fig. 1.2), where the diagonal elements contain the atomic numbers, while the off-diagonal elements specify the bond orders between the corresponding atoms in the molecular graph. Other examples include: a Cartesian distance matrix, whose off-diagonal elements are the Cartesian distances between the corresponding pairs of atoms; or a reaction network where the reactions (edges of the hypergraph) are labelled by their respective rate constants or yields.

Another very useful matrix representation of a network is the *Laplacian* matrix **L**, defined as:

$$\mathbf{L} = \mathbf{D} - \mathbf{A}, \tag{1.4}$$

where $\mathbf{D}$ is a diagonal matrix whose elements are the degrees of the nodes. Table 1.1c shows the Laplacian matrix for the same molecule (Fig. 1.2). In other words, the Laplacian matrix of a graph G is the n x n matrix $\mathbf{L}$ such that:

$$L_{ij} = \begin{cases} k_i & \text{if } i = j, \\ -A_{ij} & \text{otherwise,} \end{cases} \tag{1.5}$$

where k_i is the degree of vertex i in G (Eq. 1.3). It is apparent that any row sum of the Laplacian matrix is zero.

One can also represent a graph G by an n x m *incidence matrix* $\mathbf{M}$, whose rows correspond to the nodes of G while its columns correspond to the edges of G, as shown in Table 1.1d. The only two non-zero entries in column j of $\mathbf{M}$ correspond to the indices of the nodes incident with edge j. Likewise, the non-zero entries in row i of $\mathbf{M}$ correspond to all the edges incident to node i:

$$M_{ij} = \begin{cases} 1 & \text{if node } i \text{ is connected to edge } j, \\ 0 & \text{otherwise.} \end{cases} \tag{1.6}$$

Hence:

$$\sum_{j=1}^{M} M_{ij} = k_i. \tag{1.7}$$

We then form the matrix $\mathbf{Q}$, as shown in Table 1.1e:

$$\mathbf{Q} = \mathbf{MM}^{\mathrm{T}} \tag{1.8}$$

Thus $\mathbf{Q}$ is a symmetric n x n matrix:

$$\mathbf{Q}^{\mathrm{T}} = \left(\mathbf{MM}^{\mathrm{T}}\right)^{\mathrm{T}} = \left(\mathbf{M}^{\mathrm{T}}\right)^{\mathrm{T}}\left(\mathbf{M}^{\mathrm{T}}\right) = \mathbf{MM}^{\mathrm{T}} = \mathbf{Q} \tag{1.9}$$

with matrix elements:

$$Q_{ij} = \sum_{k=1}^{m} M_{ik}M_{kj}^{\mathrm{T}} = \sum_{k=1}^{m} M_{ik}M_{jk}. \tag{1.10}$$

For $i = j$, this yields the diagonal elements:

$$Q_{ii} = \sum_{k=1}^{m} M_{ik}M_{ik} = \sum_{k=1}^{m} M_{ik} = k_i \tag{1.11}$$

whereas for $i \neq j$, the product of the ith row with the jth row of $\mathbf{M}$ is non-zero only when M_i and M_j each have a non-zero entry in the same column, which corresponds to nodes i and j being connected to the same edge. It follows that:

$$Q_{ij} = \sum_{k=1}^{m} M_{ik}M_{kj}^{T} = A_{ij} \quad \text{for i} \neq \text{j}, \tag{1.12}$$

so that:

$$Q = D + A. \tag{1.13}$$

We can also set an arbitrary direction to each edge of G by calling one incident node of each edge the 'head' and the other the 'tail', to define an *oriented* n × m *incidence matrix* **N**, as shown in Table 1.1, with:

$$N_{ij} = \begin{cases} 1 \text{ if node i is the 'head' of edge j,} \\ -1 \text{ if node i is the 'tail' of edge j,} \\ 0 \text{ otherwise.} \end{cases} \tag{1.14}$$

Again, the only two non-zero entries in column j of **N** correspond to the nodes connected to edge j, and the k_i non-zero entries in row i of **N** correspond to all the edges connected to node i. The Laplacian matrix **L** of G can then be defined as the symmetric n x n matrix:

$$L = NN^{T} \tag{1.15}$$

with diagonal elements:

$$L_{ii} = \sum_{k=1}^{m} N_{ik}N_{ik} = \sum_{k=1}^{m} N_{ik}^{2} = k_i, \tag{1.16}$$

since there are k_i non-zero elements (each either $+ 1$ or $- 1$) in row i of **N**, whereas for i ≠ j, L_{ij} is non-zero only when N_i and N_j each have a non-zero entry in the same column, corresponding to nodes i and j being the 'head' and 'tail' of the same edge, giving the product -1 when i and j are connected. Thus the off-diagonal elements are:

$$L_{ij} = \sum_{k=1}^{m} N_{ik}N_{kj}^{T} = A_{ij} \quad \text{for i} \neq \text{j}, \tag{1.17}$$

from which Eq. (1.4) follows: $L = D - A$. So **L** and **Q** positive and semi-definite as a consequence of being products of incidence matrices.

For a weighted graph, the Laplacian matrix **L** takes the form:

$$L = WNN^{T} = D - W, \tag{1.18}$$

where $\mathbf{W}$ is the weight matrix, whose elements are the edge weights w_{ij}.

1.3 Matrix Representations of Directed and Undirected Graphs

Comparing Fig. 1.1a–c reveal one further difference the edges in the first two are non-directed, represented by simple lines, whereas the reaction network of Fig. 1.1c is a *directed hypergraph*, with the edge depicted by an arrow connecting the reactants on the left to the products on the right. Since in practice many chemical reactions are effectively irreversible under the specified experimental conditions, the relationship is not bi-directional. While an *undirected network* is represented by an adjacency matrix that is symmetric about the main diagonal, $A_{jk} = A_{kj}$, directed networks, such as the one shown in Fig. 1.3a, give rise to adjacency matrices that are not symmetric, $A_{jk} \neq A_{kj}$, as shown in Fig. 1.3b; while Fig. 1.3c depicts the weighted directed graph for a chemical reaction network, with rate constants as the edge weights, and the corresponding weighted adjacency matrix is shown in Fig. 1.3d. For a directed graph, the number of incoming edges for any node, known as the in-degree is, in general, different from the number of outgoing edges, or the out-degree. Many biological networks, such as transcription, signaling and switching networks, where specific genes or transcription factors modulate other genes, upregulating or downregulating them, form directed and weighted networks.

Table 1.2 lists several ways of classifying graphs/networks

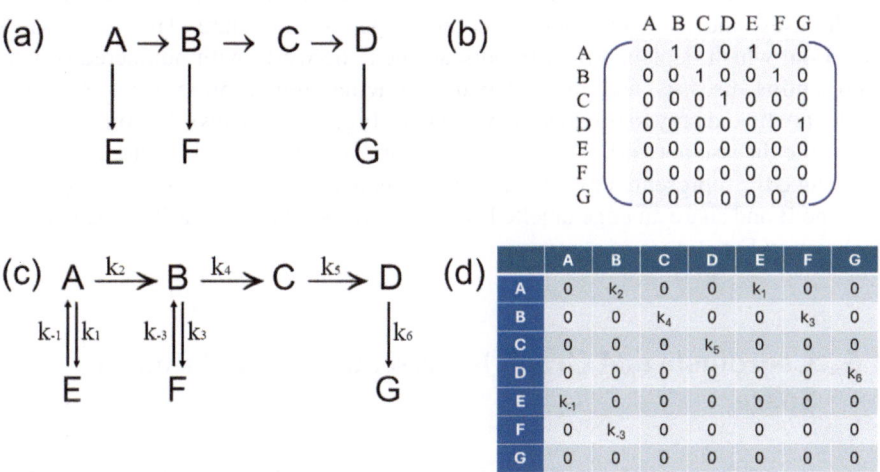

Fig. 1.3 (**a**) A directed graph and its corresponding (**b**) Adjacency matrix. (**c**) A weighted directed graph for a chemical reaction network, and its corresponding (**d**) Adjacency matrix with rate constants as the edge weights

Table 1.2 Classification of graphs/networks

Graph/network	Graph characteristics	Adjacency matrix
Undirected	Bidirectional edges; represented by lines	Symmetric $A_{ij} = A_{ji}$
Directed	Unidirectional edges; represented by arrows	Unsymmetric $A_{ij} \neq A_{ji}$
Unweighted	All edges are similar	$A_{ij} = 0$ or 1
Weighted	Edges are different	$A_{ij} =$ any real number
Unipartite	All nodes belong to the same class	Edges can connect any pair of nodes
Bipartite	Nodes belong to two distinct classes	Edges can only connect nodes of one class to nodes of the other class

1.4   Unipartite and Bipartite Networks

In a *unipartite graph* of the kind we have considered so far, all nodes belong to the same class, and any node may be connected to any other through an edge; whereas in a *bipartite graph*, the nodes belong to two distinct classes, and an edge may only connect a node of one class to a node of the other class. An example of a bipartite network is a buyer–seller network in a market, where customers can only buy from authorized sellers and not from other customers. Another example is a disease-gene association network, where specific genes are associated with specific diseases. Any bipartite graph may be converted into a unipartite graph by turning the nodes of one kind into edges. Thus if a gene is associated with multiple diseases in a disease-gene network, one can construct a new network where the nodes are all diseases, and the edges are the common genes they share. This is shown in Fig. 1.4a, which depicts a bipartite disease-gene network, with numbered squares representing diseases, and circles labelled by letters representing genes. Note that circles are linked only to squares and vice versa. Figure 1.4b shows the corresponding unipartite disease network. Diseases 1 and 2 are both regulated by gene A and thus share an edge represented by gene A. Likewise, diseases 2 and 3 are both regulated by gene B and share an edge labelled B. The corresponding unipartite gene network is shown in Fig. 1.4c.

1.5   Coordinate and Graph Representations of Chemical Space

Before we discuss similarity networks, we need to understand what we mean by molecular similarity. This is a non-trivial question, and often depends upon the specific problem at hand. Molecules can be of similar size or similar shape or similar polarity or may possess similar magnetic moment or similar electrical conductivity or may display similar anti-tumor activity. Cheminformatics algorithms generate

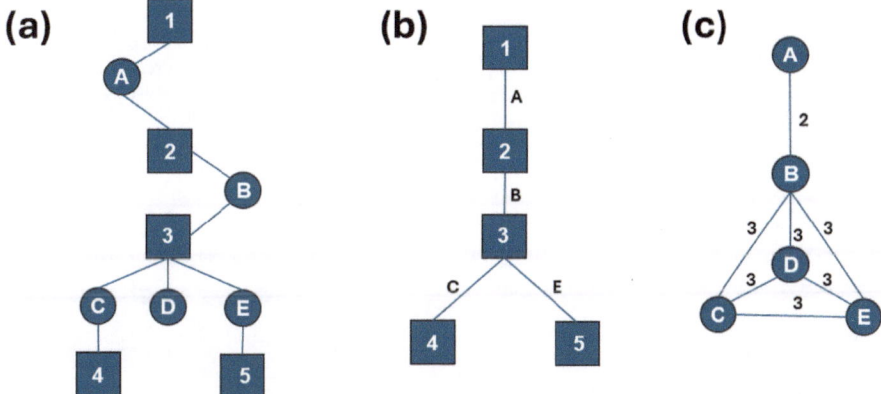

Fig. 1.4 (**a**) Schematic representation of a bipartite disease-gene network. Numbered squares represent diseases and circles (labelled by letters) represent genes. Note that circles are linked only to squares and vice versa. (**b**) The corresponding unipartite disease network. Numbered squares (nodes) represent diseases and lines (labelled by letters) represent the genes common to them. (**c**) The corresponding unipartite gene network. Labelled circles (nodes) represent genes and lines (labelled by numbers) represent the diseases common to them

many thousands of *molecular descriptors*, such as molecular weight, aqueous solubility, atom and bond counts, topological descriptors, shape descriptors, surface area descriptors, and many others, of varying levels of complexity and information content. These descriptors make up an abstract multi-dimensional space known as *chemical space*. Different regions of chemical space are populated by molecules with distinct properties, sharing some chemical similarity. Molecular similarity can be defined using a chosen set of descriptors and an appropriate similarity measure. Tens of millions of molecules are cataloged in the Beilstein database, but the chemical space of molecules obeying the rules of chemical valence has been variously estimated to be between 10^{60} and 10^{120} molecules, a number much larger than the number of atoms in the universe. High throughput screening (HTS) involves the use of automated equipment [8] or *in silico* methods such as VHTS (Virtual High Throughput Screening) to rapidly test thousands to millions of materials. Knowing the region of chemical space and the space of process parameters in which to direct one's search for new materials is of critical importance for the success of a materials discovery program. We shall have more to say on similarity measures in Chap. 2.

Instead of a coordinate-based representation of chemical space, whose coordinate axes are molecular descriptors or combinations of descriptors, as in Fig. 1.5a, an alternate, coordinate-free representation is to represent the chemical space as a similarity network, as shown in Fig. 1.5b. In a *chemical space similarity network* (CSN) [9, 10], the nodes represent distinct molecules, and the edges connect similar molecules in a pair-wise manner. All pairs of molecules with similarity (according to a pre-defined similarity measure) greater than or equal to a pre-defined similarity threshold are connected by an edge (adjacency matrix element = 1), while others

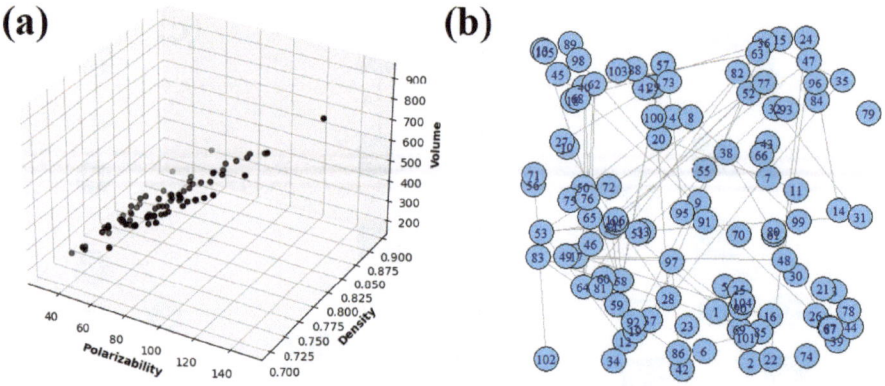

Fig. 1.5 (**a**) Coordinate-based representation of a 3-dimensional chemical space (with descriptors: polarizability, density, and volume) with 106 molecules. (**b**) Corresponding chemical space similarity network (CSN) for the same molecules

are not (adjacency matrix element = 0). Thus $A_{ij} = 1$ if $S_{ij} \geq T$ and $A_{ij} = 0$ if $S_{ij} < T$, where **A** is the adjacency matrix, **S** is the similarity matrix and T is the chosen similarity threshold. The lengths of the edges and the placement of the nodes in the graph are determined by the graph layout algorithm based on aesthetic and legibility considerations, and carry no other significance. In particular, the orientations in a similarity graph have no relation to specific descriptors or combinations thereof, unlike in a coordinate-based representation of chemical space. Of course, the calculation of molecular similarity depends upon the chosen set of molecular descriptors, as discussed above.

Different kinds of similarity metrics are appropriate for different kinds of descriptors. For descriptors consisting of binary strings, commonly known as molecular fingerprints, it is customary to employ the Tanimoto similarity index, which counts the number of bits 'ON' in both strings A and B divided by the total number of bits 'ON' in either string:

$$\tau_C = \frac{|A \cap B|}{|A \cup B|} = \frac{|A \cap B|}{|A| + |B| - |A \cap B|}. \tag{1.19}$$

Chemical similarity threshold networks depend upon the value of the chosen threshold T. Figure 1.6a–g depict the same dataset, for which the Tanimoto similarities are shown in Table 1.3a, at different similarity threshold values.

The number of edges in the network decreases with the similarity cutoff. At very low values of the similarity threshold, almost all nodes will be connected by edges, whereas at very high values of the similarity threshold, very few nodes will be connected, the adjacency matrix is sparse, and the network fragments into disconnected components, as can be seen from Table 1.4.

The Sörgel distance is defined as $1 - \tau_C$. So molecules with high similarity may be viewed as lying close together in chemical space, while those with low similarity

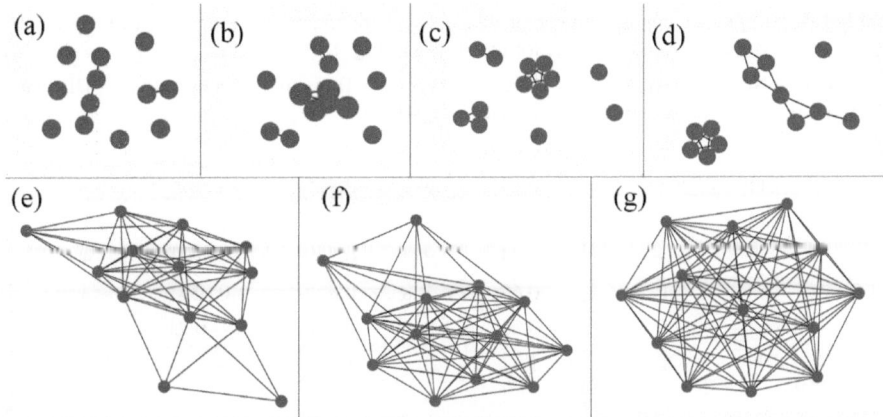

Fig. 1.6 Chemical space networks (CSN) generated from the Tanimoto similarity matrix in Table 1.3a at similarity thresholds of (**a**) 0.95 giving 4 edges, (**b**) 0.9 giving 11 edges, (**c**) 0.8 giving 14 edges, (**d**) 0.6 giving 19 edges, (**e**) 0.5 giving 54 edges, (**f**) 0.4 giving 67 edges, and (**g**) 0.3 giving a fully-connected graph with 78 edges

Table 1.3 (**a**) Tanimoto similarity matrix for a small dataset of 13 molecules, and (**b**) pIC50 values with a threshold of 12 for the same molecules

(a) Tanimoto similarity	1	2	3	4	5	6	7	8	9	10	11	12	13
1	0	0.97	0.5	0.58	0.47	0.56	0.56	0.54	0.58	0.54	0.55	0.66	0.45
2	0.97	0	0.5	0.56	0.57	0.55	0.55	0.53	0.61	0.52	0.54	0.65	0.44
3	0.5	0.5	0	0.56	0.92	0.52	0.53	0.51	0.48	0.48	0.53	0.68	0.88
4	0.58	0.56	0.56	0	0.53	0.97	0.97	0.95	0.34	0.39	0.95	0.54	0.52
5	0.47	0.57	0.92	0.53	0	0.52	0.51	0.52	0.44	0.5	0.53	0.63	0.89
6	0.56	0.55	0.52	0.97	0.52	0	0.95	0.92	0.33	0.38	0.93	0.53	0.51
7	0.56	0.55	0.53	0.97	0.51	0.95	0	0.96	0.35	0.4	0.92	0.56	0.5
8	0.54	0.53	0.51	0.95	0.52	0.92	0.96	0	0.36	0.36	0.9	0.57	0.49
9	0.58	0.61	0.48	0.34	0.44	0.33	0.35	0.36	0	0.55	0.32	0.4	0.43
10	0.54	0.52	0.48	0.39	0.5	0.38	0.4	0.36	0.55	0	0.38	0.51	0.45
11	0.55	0.54	0.53	0.95	0.53	0.93	0.92	0.9	0.32	0.38	0	0.52	0.55
12	0.66	0.65	0.68	0.54	0.63	0.53	0.56	0.57	0.4	0.51	0.52	0	0.59
13	0.45	0.44	0.88	0.52	0.89	0.51	0.5	0.49	0.43	0.45	0.55	0.59	0
(b) Molecule	1	2	3	4	5	6	7	8	9	10	11	12	13
PIC50	7.49	7.2	8.05	5.32	7.32	7.62	5.93	6.62	4.77	4.8	6.83	3.32	4.23

Table 1.4 Number of edges, number of connected components, edge density, average degree, transitivity, assortativity and diameter for the graphs in Fig. 1.6

Similarity Threshold	Number of edges	Number of components	Edge density	Average degree	Transitivity	Assortativity	Diameter
0.95	4	9	0.05	0.615	0	0	3
0.9	11	7	0.14	1.692	0.875	0.78	2
0.8	14	6	0.18	2.154	1	1	1
0.6	19	3	0.24	2.923	0.830	0.404	4
0.5	54	1	0.69	8.308	0.828	0.034	3
0.4	67	1	0.86	10.308	0.906	− 0.218	2
0.3	78	1	1	12	1	–	1

are separated by greater Sörgel distance. Another related similarity measure is the cosine similarity:

$$S_C(a, b) = \frac{\mathbf{a.b}}{\|\mathbf{a}\| \|\mathbf{b}\|}, \tag{1.20}$$

where $\mathbf{a}$ and $\mathbf{b}$ are descriptor vectors for the two molecules, and $\|\mathbf{a}\|$ and $\|\mathbf{b}\|$ are their respective norms. In the next chapter, we will deal with nonlinear generalizations of this similarity metric.

When molecules are represented by character strings, as in amino acid sequences, the distance between two sequences is given by the Levenshtein distance or edit distance, which is the minimum number of edits (substitutions, single-character insertions and deletions) required to convert one string into the other. For strings of equal length, this reduces to the Hamming distance which is simply the minimum number of substitutions required to convert one string into the other. With real valued descriptors, the most commonly employed distance measure is the Euclidean distance:

$$R_{ab} \sqrt{\sum_i (a_i - b_i)^2}, \tag{1.21}$$

where a_i and b_i are the values of the ith descriptor for the two molecules a and b.

1.6   Feature Networks

As mentioned in the previous section, many thousands of molecular descriptors are available through common cheminformatics algorithms, and these make up a chemical space of multiple dimensions. Which of these descriptors are most relevant for description of a particular problem is therefore a non-trivial question. Often some or many of these descriptors are inter-related, but not identical. These molecular

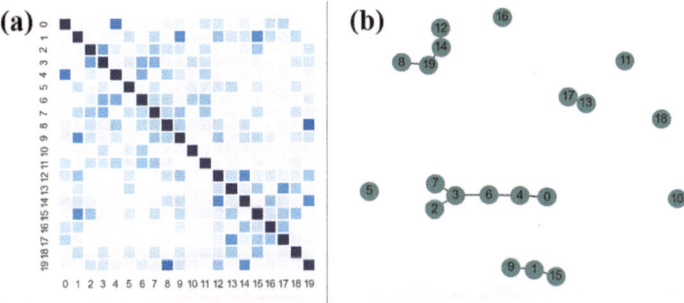

Fig. 1.7 (**a**) Feature correlation matrix as a heat map. (**b**) Network representation of the same features

descriptors or features may themselves be treated as an abstract network. Thus, as distinct from the various kinds of molecular networks discussed above, we can also construct a (unipartite, undirected) feature network, where the nodes are molecular descriptors or features, and the edges represent the correlations between them. The correlation matrix is a natural matrix representation of a feature network, where the edge weights are the correlation coefficients between the respective features, to form a weighted feature network. One can also construct an unweighted feature network by representing only the most inter-correlated pairs of features with edges, after setting a correlation threshold. Figure 1.7a shows the feature correlation matrix between 20 descriptors in a dataset, in the form of a heatmap, and Fig. 1.7b shows the unweighted network constructed from this network after applying a correlation threshold. The descriptors here are seen to consist of a few correlated clusters and a few singletons uncorrelated with any of the others.

Such correlation analysis is useful in *feature selection*—choosing a subset of independent features for machine learning (ML) [11]. In the absence of feature selection, many ML algorithms suffer from the so-called *curse of dimensionality*, which is particularly acute for small datasets where the number of training molecules is vastly exceeded by the number of descriptors in the model. Redundant descriptors introduce free parameters and noise in the model, resulting in a greater tendency for *overfitting*—the tendency of a model to "memorize" specific features of the molecules used for training, resulting in loss of generalization ability and poor performance on hitherto unseen molecules, in spite of excellent performance on the training set. As we will discuss in Sect. 6.6, overfitting is a result of poorly sampled directions in feature space. Furthermore, as the number of dimensions increases, the volume of the space grows exponentially, resulting in the data becoming increasingly sparse, and the concept of distance in this multi-dimensional space becomes less discriminative [12], posing problems for many ML algorithms. We then require very large datasets to model such high-dimensional data. Some feature selection techniques will be discussed in later chapters.

There are two main sources of prediction error in any model: bias and variance. The difference between the average prediction of a model and the true value is its

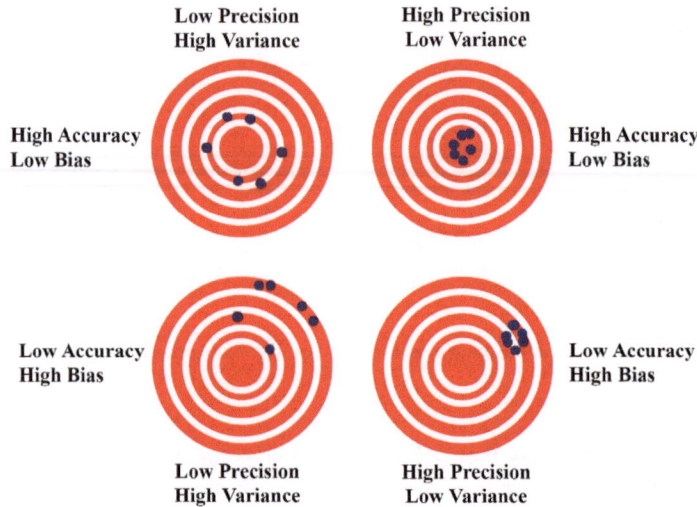

Fig. 1.8 Bulls-eye diagram illustrating bias and variance in a model's predictions

bias, while variance is the variability in the model prediction for any given data point, as illustrated in Fig. 1.8. In traditional machine learning, there is a trade-off between the two [13]. Increasing the number of parameters in a model reduces the bias, but at the cost of increasing the variance in the predictions. But as we shall see in Sect. 6.6, over-parameterized deep learning (DL) models defy this traditional wisdom, allowing such models to achieve zero training error without loss of generalization ability [14, 15].

References

1. W. Blake, Auguries of innocence, in *The Complete Poetry and Prose of William Blake*, eds. by D.V. Erdman (Newly revised ed., Anchor Books, 1988), p. 490. ISBN: 0385152132
2. M.E. Newman, The mathematics of networks. New Palgrave Encyclopedia Econ. **2**, 1–12 (2008)
3. M.P. Krein, N. Sukumar, Exploration of the topology of chemical spaces with network measures. J. Phys. Chem. A **115**(45), 12905–12918 (2011). https://doi.org/10.1021/jp204022u
4. H.C. Longuet-Higgins, M.D.V. Roberts, Proc. R. Soc. A **224**, 336–347 (1954)
5. W.N. Lipscomb, Acc. Chem. Res. **6**, 257–262 (1973)
6. D. Weininger, SMILES, a chemical language and information system. 1. Introduction to methodology and encoding rules. J. Chem. Inf. Comput. Sci. **28**, 31–36 (1988). https://doi.org/10.1021/ci00057a005
7. D. Weininger, A. Weininger, J.L. Weininger, SMILES. 2. Algorithm for generation of unique SMILES notation. J. Chem. Inf. Comput. Sci. **29**, 97–101 (1989). https://doi.org/10.1021/ci00062a008
8. N.J. Szymanski, B. Rendy, Y. Fei, R.E. Kumar, T. He, D. Milsted, M.J. McDermott, M. Gallant, E.D. Cubuk, A. Merchant, H. Kim, A. Jain, C.J. Bartel, K. Persson, Y. Zeng, G. Ceder, An

autonomous laboratory for the accelerated synthesis of novel materials. Nature **624**, 86–91 (2023). https://doi.org/10.1038/s41586-023-06734-w

9. G.M. Maggiora, J. Bajorath, Chemical space networks: a powerful new paradigm for the description of chemical space. J. Comput. Aided Mol. Des.Comput. Aided Mol. Des. **28**(8), 795–802 (2014). https://doi.org/10.1007/s10822-014-9760-0
10. N. Sukumar, M.P. Krein, Graphs and networks in chemical and biological informatics: past, present, and future. Future Med. Chem. **4**(16) (2012). https://doi.org/10.4155/fmc.12.128
11. N. Sukumar, H. Anandaram, P. Bhadra, *Computational Drug Discovery—A Primer* (Ion Cures Press, 2023)
12. M. Wolf, The math behind "the curse of dimensionality", in *Towards Data Science*. https://towardsdatascience.com/the-math-behind-the-curse-of-dimensionality-cf8780307d74
13. S. Fortmann-Roe, *Understanding the Bias-Variance Tradeoff* (June 2012). http://scott.fortmann-roe.com/docs/BiasVariance.html
14. M. Belkin, Fit without fear: remarkable mathematical phenomena of deep learning through the prism of interpolation. arXiv:2105.14368v1 (2021)
15. J.W. Rocks, P. Mehta, Memorizing without overfitting: bias, variance, and interpolation in overparameterized models. Phys. Rev. Res. **4**, 013201 (2022). https://doi.org/10.1103/PhysRevResearch.4.013201

Chapter 2
Transformations of Chemical Space

The universe is transformation; life is opinion.
—*Marcus Aurelius* [1]

2.1 Vector Spaces and Metric Tensors

In this chapter, we will introduce tangent vectors and similarity kernels, two key concepts upon which much of the discussion in later chapters will be based. Consider a general chemical space in N dimensions $\mathbb{R}^N$, where the coordinate axes are continuous real-valued molecular descriptors (or combinations of descriptors). Note that any arbitrary point in the space need not correspond to a chemically realizable molecule. In the case where the descriptors are orthogonal, we have a simple Euclidean space. If the descriptors are continuous and real-valued, as we have assumed, the space forms a real *vector space*, with the following properties defined on it, with **u**, **v** and **w** being elements of the vector space, and a and b being scalars:

- Vector addition obeys the associative property:

$$(\mathbf{u} + \mathbf{v}) + \mathbf{w} = \mathbf{u} + (\mathbf{v} + \mathbf{w}) \tag{2.1}$$

- Vector addition obeys the commutative property:

$$\mathbf{u} + \mathbf{v} = \mathbf{v} + \mathbf{u} \tag{2.2}$$

- An identity element (**0**) with respect to vector addition exists, such that:

$$\mathbf{v} + \mathbf{0} = \mathbf{v} \tag{2.3}$$

© The Author(s), under exclusive license to Springer Nature Switzerland AG 2024
N. Sukumar, *Navigating Molecular Networks*,
SpringerBriefs in Materials, https://doi.org/10.1007/978-3-031-76290-1_2

- An inverse element $(-\mathbf{v})$ with respect to vector addition exists for any vector $\mathbf{v}$, such that:

$$\mathbf{v} + (-\mathbf{v}) = \mathbf{0} \tag{2.4}$$

- Scalar multiplication is compatible with field multiplication, i.e.:

$$a(b\mathbf{v}) = (ab)\mathbf{v} \tag{2.5}$$

- An identity element 1 with respect to scalar multiplication exists, i.e.:

$$1\mathbf{v} = \mathbf{v} \tag{2.6}$$

- Scalar multiplication obeys the distributive property with respect to vector addition:

$$a(\mathbf{u} + \mathbf{v}) = a\mathbf{u} + a\mathbf{v} \tag{2.7}$$

- Scalar multiplication obeys the distributive property with respect to field addition:

$$(a + b)\mathbf{v} = a\mathbf{v} + b\mathbf{v}. \tag{2.8}$$

A set of linear independent vectors that spans the entire vector space is called a *basis*. Any vector in this space can be represented as a linear combination of the basis vectors. A *linear transformation* $\mathbf{A}$ that transforms vectors $\mathbf{u}$ and $\mathbf{v}$ to vectors $\mathbf{u}' = \mathbf{Au}$ and $\mathbf{v}' = \mathbf{Av}$, respectively, satisfies two properties:

$$\mathbf{A}(\mathbf{u} + \mathbf{v}) = \mathbf{A}(\mathbf{u}) + \mathbf{A}(\mathbf{v}) \tag{2.9}$$

$$\mathbf{A}(c\mathbf{u}) = c\mathbf{A}(\mathbf{u}), \text{ where } c \text{ is a real or complex scalar.} \tag{2.10}$$

The *inner product* (or dot product) of vectors $\mathbf{u}$ and $\mathbf{v}$ is defined as:

$$\langle \mathbf{u}|\mathbf{v} \rangle \equiv \mathbf{u} \cdot \mathbf{v} \equiv \mathbf{u}^{\mathrm{T}}\mathbf{v}, \tag{2.11}$$

where $\mathbf{u}^{\mathrm{T}}$ is the transpose of $\mathbf{u}$. If the two vectors $\mathbf{u}$ and $\mathbf{v}$ represent molecules in the chemical space, then the inner product $\langle \mathbf{u}|\mathbf{v} \rangle$ is a linear measure of the chemical similarity between the molecules, and is related the distance $||\mathbf{u} - \mathbf{v}||$ between $\mathbf{u}$ and $\mathbf{v}$ in the chemical space through:

$$||\mathbf{u} - \mathbf{v}|| = \sqrt{(\mathbf{u} \cdot \mathbf{u} + \mathbf{v} \cdot \mathbf{v} - 2\mathbf{u} \cdot \mathbf{v})}. \tag{2.12}$$

If $\prec u|v \succ = 0$, $\mathbf{u}$ and $\mathbf{v}$ are said to be orthogonal. The inner products $\mathbf{u} \cdot \mathbf{u}$ and $\mathbf{v} \cdot \mathbf{v}$ are called the norm of the vectors $\mathbf{u}$ and $\mathbf{v}$, respectively. If the norm is unity, the vector is said to be normalized, as required for its interpretation as a self-similarity.

The concept of molecular similarity underlies all Quantitative Structure Activity/Property Relationship (QSAR/QSPR) modeling. This is because molecular structure, or the relative positions of the atoms in a molecule, determines the molecular Hamiltonian, whose solutions in turn determine all molecular properties. In the field of QSAR, the *Similarity Principle* embodies the expectation that structurally similar molecules will have similar biological activities, while dissimilar molecules will have dissimilar biological activities. This is the basis of ligand-based drug design, a computational drug design strategy undertaken without knowledge of the structure of the biological target (protein or nucleic acid), by exploiting similarities between molecules ("ligands") known to produce the desired biological activity by modulating the target. As we will see in Sect. 5.4, however, there are several exceptions to the Similarity Principle, giving rise to rugged, non-differentiable structure–activity landscapes and Activity Cliffs [2], which are pairs of structurally similar molecules with very different biological activities.

As long as we restrict ourselves to a differentiable manifold M, we can associate with every point $p \in M$ of such a manifold, a real vector space T_pM, called the tangent space of M at p, that contains the possible directions of the tangent vectors passing through p. The dimension of the tangent space at every point of a connected manifold is equal to that of the manifold itself. The *metric tensor*, defined on a manifold, allows for the definition of lengths and angles. An example is the *Riemannian metric* tensor, defined on a Riemannian manifold, that makes it possible to define geometric notions, such as angles, lengths of curves, areas of surfaces, volumes and other higher-dimensional objects, as well as curvatures of submanifolds and of the manifold itself.

A Riemannian metric g assigns a positive-definite inner product g_p to each point p on the manifold M, that maps the tangent vector space to the field of scalars: $T_pM \otimes T_pM \rightarrow R$. This induces a norm $|\cdot|_p: T_pM \rightarrow R$:

$$|v|_p = \sqrt{g_p(v, v)}. \tag{2.13}$$

The Riemannian metric can be written as:

$$g = \sum_{i,j} g_{ij} \, dx^i \otimes dx^j. \tag{2.14}$$

and its components at any point p are:

$$g_{ij}|_p = g_p \left(\frac{\partial}{\partial x_i|_p}, \frac{\partial}{\partial x_j|_p} \right). \tag{2.15}$$

The related *Ricci curvature* is a symmetric tensor that quantifies the extent of deformation of the space from a locally Euclidean geometry as one moves along geodesics in the space. We will return to the Ricci curvature tensor in Sect. 3.5.

2.2   Dimensionality Reduction

The simplest kind of transformation we can perform on chemical space is a linear transform, effected by a linear operator $\mathbf{A}$:

$$\mathbf{v'} = \mathbf{Av} \tag{2.16}$$

for all vectors $\mathbf{v}$ in the vector space. If both $\mathbf{v}, \mathbf{v'} \in \mathbb{R}^N$, then $\mathbf{A}$ is a square matrix of dimensions $N \times N$. In the special case where $\mathbf{A}$ is an orthogonal matrix, the transformation is norm preserving, and represents a simple rotation of the coordinate axes.

An example of such a transformation is the popular multivariate data analysis technique, *Principal Component Analysis* (PCA) [3], employed for data visualization and dimensionality reduction in machine learning. Principal components are linear transformations of the original data vectors $\mathbf{v_i} \in \mathbb{R}^N$ to $\mathbf{PC_j} \in \mathbb{R}^M$ ($M \leq N$), ordered such that the first principal component (PC1) accounts for the maximum variance in the dataset, with each subsequent component accounting for decreasing amounts of the residual variance, and such that all principal components are orthogonal to one other.

$$
\begin{aligned}
PC_1 &= c_{11}v_1 + c_{12}v_2 + \cdots + c_{1N}v_N \\
PC_2 &= c_{21}v_1 + c_{22}v_2 + \cdots + c_{2N}v_N \\
PC_3 &= c_{31}v_1 + c_{32}v_2 + \cdots + c_{3N}v_N \\
&\quad \cdots \qquad \cdots \\
PC_M &= c_{M1}v_1 + c_{M2}v_2 + \cdots + c_{MN}v_N
\end{aligned}
\tag{2.17}
$$

or in matrix notation:

$$\mathbf{PC} = \mathbf{c}^T\mathbf{v}. \tag{2.17a}$$

The matrix $\mathbf{c}$, of dimension $M \times N$, known as the Loadings Matrix, contains the contributions ("loadings") of the original descriptors to the principal components. The complementary "scores" matrix, of dimension $P \times M$, describes the P molecules in terms of the M principal components. What has been effected by the transformation is a compact *encoding* of the original data in $\mathbb{R}^N$ to a new vector space $\mathbb{R}^M$ of lower dimensionality $M \leq N$, while preserving its essential features. The maximum number of principal components M that can be extracted from a dataset is the smaller of the number of original descriptors N and the number of molecules P in the dataset.

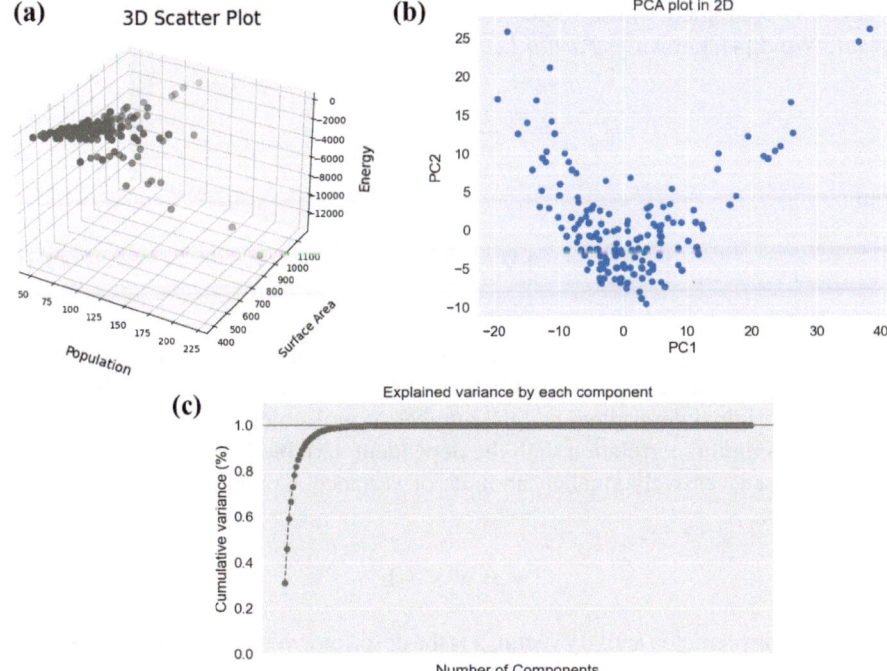

Fig. 2.1 (**a**) A dataset of 278 molecules described by 248 descriptors (only 3 of which are shown) and (**b**) its encoding onto a space of the first two principal components (PC1 and PC2) that capture the major part of the variance in the original dataset, as shown in (**c**)

Figure 2.1 shows such a principal component encoding of a multi-dimensional dataset onto the space of its first two principal components, capturing the major part of the variance in the original dataset.

In most applications we are interested in some specific aspects of molecular similarity. For instance, in classifying materials into conductors or insulators, or into soluble or insoluble/sparingly soluble in a given solvent, the color might be quite irrelevant. It is then useful to project the chemical space onto a lower dimensional space containing only those descriptors or combinations of descriptors relevant to the property of interest. Dimensionality reduction is behind many commonly used visualization methods in machine learning such as PCA, t-distributed Stochastic Neighbor Embedding (t-SNE), Uniform Manifold Approximation and Projection (UMAP) and Self Organized Maps (SOM, also known as Kohonen maps after their inventor Teuvo Kalevi Kohonen). We will discuss these maps in more detail in Chap. 4.

PCA is combined with a regression step in an iterative procedure, devised by Svante Wold [4], known as *Partial Least Squares* (PLS) regression:

$$y = \sum_i c_i LV_i + b, \tag{2.18}$$

with:

$$LV_i = \sum_j a_{ij} x_j, \tag{2.19}$$

where the c_i are regression coefficients. As in PCA, the latent variables (LV) are orthogonal to each other, and the number of latent variables is the smaller of the number of original descriptors and the number of molecules in the dataset, but now LV1 has maximum correlation with the dependent variable (y), and successive LVs account for successively smaller amounts of variance. Combining Eqs. (2.18) and (2.19) gives:

$$y = \mathbf{w}^T\mathbf{x} + \mathbf{b}, \tag{2.20}$$

where $\mathbf{y}$ is the predicted activity vector, $\mathbf{x}$ is the descriptor vector, $\mathbf{b}$ is a bias vector, and $\mathbf{w}$ is a weight matrix that combines the regression coefficients and the LV loadings:

$$w_{ij} = \sum_i c_i a_{ij}. \tag{2.21}$$

Linear regression is performed with the function:

$$f(\mathbf{w}, \mathbf{x}) = \mathbf{w}^T\mathbf{x} + \mathbf{b}, \tag{2.22}$$

and a squared loss function $L(w)$ that expresses the squared deviation of the predicted function value $f(w, x)$ from the actual value of y:

$$L(\mathbf{w}) = \frac{1}{2} \sum_i \{f(\mathbf{w}, x_i) - y_i\}^2. \tag{2.23}$$

The objective is to minimize the loss $L(w)$ by varying the model parameters w. Regression is performed with gradient descent, so that the weights at any step t are given by:

$$\mathbf{w}^{(t+1)} = \mathbf{w}^{(t)} - \eta_t \, \nabla_w L\left(\mathbf{w}^{(t)}\right), \tag{2.24}$$

where η_t is the learning rate. The second term is the gradient of the loss function:

$$\nabla_w L\big(w^{(t)}\big) = \frac{dL\big(w^{(t)}\big)}{dw} = \sum_i \big\{f(w, x_i) - y_i\big\} \nabla_w f(w, x_i). \qquad (2.25)$$

We note from Eq. (2.22), that the model is linear in the inputs, so that:

$$\nabla_w f(w, x) = \frac{df(w, x)}{dw} = x. \qquad (2.26)$$

2.3 Similarity Kernels and Kernel Methods

We can also consider a transformation that maps the original data onto a vector space $\mathbb{R}^M$ of higher dimensionality $M > N$. We then have:

$$\langle u'|v'\rangle = \langle Au|Av\rangle = u^T A^T A v = u^T K v \equiv \langle u|K|v\rangle, \qquad (2.27)$$

where

$$K = A^T A \qquad (2.28)$$

is known a *kernel* function (not to be confused with the kernel of a linear map). A kernel is a positive semi-definite symmetric function of two inputs that describes the similarity between them. The kernel function K here defines a nonlinear similarity measure between the original vectors v and u:

$$\langle u|K|v\rangle \equiv u^T K v. \qquad (2.29)$$

If K is the identity matrix, this reduces to the linear similarity measure (2.11).

Molecular similarity assessment underlies all QSAR/QSPR modeling, by virtue of the similarity principle mentioned above. Therefore the similarity kernels that are of most practical interest are those that preserve the important similarity relationships between molecules. A similarity measure determines the distance between molecules in chemical space, with similar molecules lying closer together and dissimilar ones farther apart. Isometric transformations are those that preserve distances exactly, but this is too restrictive for most practical purposes. We are interested instead in transformations where molecules lying close together in the original space $\mathbb{R}^N$ are still relatively close in the higher dimensional space $\mathbb{R}^M$ and molecules lying farther apart in $\mathbb{R}^N$ are also farther apart in $\mathbb{R}^M$.

A commonly used nonlinear similarity kernel is the radial basis function (RBF) kernel, obtained with a Gaussian form for $\mathbf{K}$:

$$\langle \mathbf{u} | \mathbf{K}^{\mathbf{RBF}} | \mathbf{v} \rangle = \exp\left\{ -\frac{(\mathbf{u}_i - \mathbf{v}_i)^2}{\sigma} \right\}, \tag{2.30}$$

where $\mathbf{u}$ and $\mathbf{v}$ are descriptor vectors for the two molecules, and σ is a parameter specifying the width of the Gaussian function. This kernel maps the nonlinear similarity measure in $\mathbb{R}^N$ to a linear cosine similarity in $\mathbb{R}^M$. This is known as the kernel trick, whereby a complicated nonlinear similarity relationship between molecules in the original descriptor space is transformed into a dot product relationship in the higher dimensional space $\mathbb{R}^M$.

Support Vector Machines (SVM), introduced by Vapnik [5], provide a nice illustration of the kernel trick, as well as the concept of Lagrangian duality. In SVM classification one seeks a hypersurface in chemical space that effects the optimum separation between two classes, say active drugs with class label $y_i = 1$, and inactive molecules with class label $y_i = -1$. If the two classes are linearly separable, there exists an optimum hyperplane:

$$\left[\mathbf{w}^T \mathbf{x}\right] + \mathbf{b} = 0 \tag{2.31}$$

separating them. Here $\mathbf{x}$ is a vector representing a molecule in a multi-dimensional descriptor space. Among the various possible separating hyperplanes, the optimal one is that which maximizes the margin or separation between two auxiliary hyperplanes, shown in Fig. 2.2a, defined by:

$$\left[\mathbf{w}^T \mathbf{x}\right] + \mathbf{b} = \pm 1. \tag{2.32}$$

The positive class is defined by $\left[\mathbf{w}^T \mathbf{x}\right] + \mathbf{b} > 1$, and the negative class by $\left[\mathbf{w}^T \mathbf{x}\right] + \mathbf{b} < -1$. Maximizing the margin in feature space minimizes the risk of overfitting, and thus yields the best separating hyperplane. Molecules lying on the auxiliary hyperplanes are called support vectors. The margin is given by:

$$\text{margin} = \frac{2}{\|\mathbf{w}\|^2} \tag{2.33}$$

The Lagrangian for this constrained optimization problem, namely that of maximizing the margin or minimizing $\frac{1}{2}\|\mathbf{w}\|^2$ is:

$$\mathbf{L} = \frac{\|\mathbf{w}\|^2}{2} - \sum_{i=1}^{N} \alpha_i \left[y_i \left(\mathbf{w}^T \mathbf{x} + \mathbf{b} \right) - 1 \right], \tag{2.34}$$

where the α_i are Lagrange multipliers for the constraints:

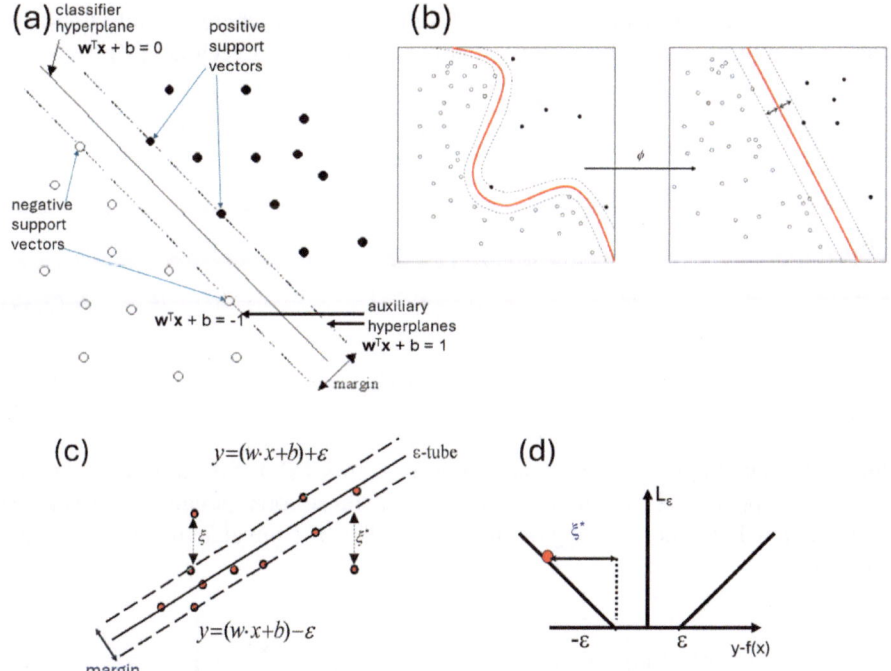

Fig. 2.2 (**a**) SVM classification, showing positive and negative support vectors, classifier and auxiliary hyperplanes. The dashed orange lines show alternative classifiers. (**b**) Kernel trick to transform a non-linearly separable function into a higher dimensional linearly separable function. (**c**) Support Vector Regression (SVR). Overfitting is avoided by minimizing the regularized empirical error and controlling the model complexity. (**d**) ε-insensitive loss function $L_\varepsilon = 0$ if $|y - f(x)| \le \varepsilon$, $L_\varepsilon = |y\text{-}f(x)| - \varepsilon$ otherwise. Adapted with permission from [6] © 2023, Ion Cures Press. All rights reserved

$$y_i \left(\left[\mathbf{w}^T \mathbf{x}_i \right] + \mathbf{b} \right) \ge 1. \tag{2.35}$$

Differentiation of the Lagrangian (2.34) with respect to **w** and **b** gives:

$$\partial L / \partial \mathbf{w} = 0 \Rightarrow \mathbf{w} = \sum_{i=1}^{N} \alpha_i y_i x_i, \tag{2.36}$$

$$\partial L / \partial \mathbf{b} = 0 \Rightarrow \sum_{i=1}^{N} \alpha_i y_i = 0. \tag{2.37}$$

In practice, SVM involves solution of a dual problem where the only non-zero Lagrange multipliers are those corresponding to the support vectors (SV):

$$\max_{\alpha}(w_\alpha) = \max_{\alpha} \left\{ \sum_{i}^{SV} \alpha_i - \frac{1}{2} \sum_{i}^{SV} \sum_{j}^{SV} \alpha_i \alpha_j y_i y_j x_i^T x_j \right\}, \tag{2.38}$$

with:

$$\alpha_i \geq 0 \tag{2.39}$$

$$\sum_i \alpha_i y_i = 0, \tag{2.40}$$

where the summations run over all support vectors (SV). So the only Lagrange multipliers appearing in the summations above are those corresponding to the support vectors, namely the points lying on the margin. This is equivalent to solving for the optimal α^*:

$$\alpha^* = \arg \min_{\alpha} \left\{ \frac{1}{2} \sum_{i}^{SV} \sum_{j}^{SV} \alpha_i \alpha_j y_i y_j x_i^T x_j - \sum_{i}^{N} \alpha_i \right\}. \tag{2.41}$$

The equation for the optimal separating hyperplane is then given by:

$$\mathbf{w}^* = \sum_{i}^{SV} \alpha_i y_i x_i \tag{2.42}$$

$$\mathbf{b}^* = \frac{1}{2} (\mathbf{w}^*)^T \sum_{i}^{SV} x_i. \tag{2.43}$$

Note that only the support vectors appear in the Eqs. (2.42)–(2.43) defining the parameters $\mathbf{w}^*$ and $\mathbf{b}^*$ of the model. The class prediction for a test point $\mathbf{x}$ is then given by:

$$f(\mathbf{w},\ \mathbf{x}) = \text{sign}\left[(\mathbf{w}^*)^T \mathbf{x} + \mathbf{b}^* \right]. \tag{2.44}$$

Overfitting is controlled through a user-defined regularization parameter C that determines the trade-off between margin maximization and misclassification, such that:

$$0 \leq \alpha_i \leq C. \tag{2.45}$$

This results in soft-margin classification.

When the molecules cannot be separated by a linear partition in the original chemical space, the kernel trick is employed to map the data onto a higher-dimensional space $\varphi: \mathbb{R}^N \to \mathbb{R}^M$ (M > N), as shown in Fig. 2.2b, with an appropriately chosen kernel:

$$K(x_i, x_j) \equiv \langle \varphi(x_i) | \varphi(x_j) \rangle \tag{2.46}$$

such that the molecules belonging to different classes are linearly separable. The dual problem is then:

$$\max_{\alpha}(W_\alpha) = \max_{\alpha} \left\{ \sum_i^{SV} \alpha_i - \frac{1}{2} \sum_i^{SV} \sum_j^{SV} \alpha_i \alpha_j y_i y_j \langle \varphi(x_i) | \varphi(x_j) \rangle \right\} \tag{2.47}$$

$$\alpha^* = \arg\min_{\alpha} \left\{ \frac{1}{2} \sum_i^{SV} \sum_j^{SV} \alpha_i \alpha_j y_i y_j K(x_j, x_i) - \sum_i^{N} \alpha_i \right\}. \tag{2.48}$$

Introduction of the kernel function enables computations to be performed in the input space itself.

When SVM is used for regression, it is called *support vector regression* (SVR), to fit an equation of the form:

$$\left[w^T x \right] + b = y. \tag{2.49}$$

As before, we construct auxiliary hyperplanes, but now the objective is to accommodate as many of the data points as possible within the auxiliary hyperplanes, and to minimize the margin, as shown in Fig. 2.2c. Overfitting is avoided by minimizing the regularized empirical error, which is defined as the sum of the training error and the model complexity, rather than minimizing the training error itself:

$$\min \left\{ C \sum_i (\xi_i + \xi_i^*) + \frac{1}{2} ||w||^2 \right\}, \tag{2.50}$$

subject to:

$$y_i - w^T x_i - b \leq \varepsilon + \xi_i$$
$$w^T x_i + b - y_i \leq \varepsilon + \xi_i^* \tag{2.51}$$
$$\xi_i, \xi_i^* \geq 0$$

and controlling the model capacity C. Here ξ_i and ξ_i^* are the distances of the data points from the two auxiliary hyperplanes. The ε-insensitive loss function introduced by Vapnik, and shown schematically in Fig. 2.2d:

$$L_\epsilon(y-f(x)) = \min(0, |y-f(x)|-\epsilon) \tag{2.52}$$

ensures that the model is penalized only for data points falling outside the margin. For nonlinear problems, SVR again employs the kernel trick to perform linear regression in the high-dimensional feature space, so that we now have, instead of Eq. (2.49):

$$\left[\mathbf{w}^\mathsf{T}\varphi(\mathbf{x})\right] + \mathbf{b} = \mathbf{y}. \tag{2.53}$$

The kernel trick is likewise employed to achieve a nonlinear generalization of PLS, namely *Kernel Partial Least Squares* (KPLS) regression, where Eq. (2.20) is now replaced by:

$$\mathbf{y} = \mathbf{w}^\mathsf{T}\mathbf{K}\mathbf{x} + \mathbf{b}, \tag{2.54}$$

where $\mathbf{K}$ is a nonlinear kernel matrix. We perform a kernel regression with the function:

$$f(\mathbf{w}, \mathbf{x}) = \mathbf{w}^\mathsf{T}\mathbf{K}\mathbf{x} + \mathbf{b} \tag{2.55}$$

by minimizing the squared loss function L(w) given by Eq. (2.23) using gradient descent. The gradient of the loss function is still given by Eq. (2.25), but now by virtue of Eq. (2.55), we have:

$$\nabla_\mathbf{w}f(\mathbf{w}, \mathbf{x}) = \mathbf{K}\mathbf{x}. \tag{2.56}$$

The function f(w, x) is still linear in the model parameters $\mathbf{w}$, because the kernel $\mathbf{K}$ does not depend on $\mathbf{w}$, although it is nonlinear in the inputs $\mathbf{x}$.

Gaussian Process Regression (GPR), also known as *Kriging*, is a probabilistic regression method based on the assumption that the function to be predicted is drawn from a Gaussian process [7–9]. This enables uncertainty quantification along with the predictions, providing a confidence estimate for the predicted function. GPR is a form of Bayesian optimization, enabling predictions to be made about the data by incorporating prior knowledge. There can be infinitely many functions that can fit any given set of training data. A Gaussian process specifies a probability distribution over the possible functions. GPR seeks to find a distribution over the possible functions f(x) that are consistent with the training data. One begins with a prior distribution and updates it as training data points are added, to generate the posterior distribution over functions. In a multivariate distribution, each random variable follows a Gaussian distribution, as does their joint distribution. This multivariate Gaussian distribution is specified by a mean and a covariance matrix. The covariance matrix is necessarily symmetric and positive semi-definite, and it determines which functions from the space of all possible functions are more probable. The diagonal elements of the covariance matrix are the variances of the random variables, and its off-diagonal elements describe correlations between these variables. If $\mathbf{X}$ is the column vector:

$$\mathbf{X} = (X_1, X_2, \ldots, X_n)^{\mathrm{T}}, \tag{2.57}$$

then the covariance matrix $\mathbf{K}$ is the matrix with elements:

$$K(X_i, X_j) = \frac{1}{n} \sum_{i,j=1}^{n} [(X_i - \overline{X_i})(X_j - \overline{X_j})], \tag{2.58}$$

where $\overline{X_i}$ and $\overline{X_j}$ are the mean values of X_i of X_j respectively. If the $\mathbf{X}$ variables are mean-centered, then $\mathbf{K}$ is simply:

$$\mathbf{K} = \frac{\mathbf{X}^{\mathrm{T}}\mathbf{X}}{n}. \tag{2.58a}$$

The kernel of the Gaussian process is specified by the elements of the covariance matrix, and consists of:

$$\begin{pmatrix} K_{\text{train}} & K_{\text{train}-\text{test}} \\ K_{\text{train}-\text{test}}^{\mathrm{T}} & K_{\text{test}} \end{pmatrix}. \tag{2.59}$$

K_{train} contains the similarities of the training molecules to each other, K_{test} contains the similarities of the test molecules to each other, and $K_{\text{train-test}}$ contains the similarities of the training molecules to the test molecules. The kernel is thus a similarity network (represented in general as a weighted network). The covariance matrix is used to ensure that the function is not too rough, so that two molecules that are estimated to be similar by the kernel get mapped to values that are close together in the output space. GP regression is performed on the training data by Bayesian inference, using the conditional probability of the test data given the training data: $P(f_{\text{test}}|x_{\text{test}}, f_{\text{train}}, x_{\text{train}})$—this also follows a Gaussian distribution—and updating the probabilities of the outputs for the test molecules with each new training molecule. The mean of the conditional or posterior distribution gives the most probable regression fit for the data. The training data points constrain the set of possible functions to only those that pass through the training points in the absence of added noise, as shown in Fig. 2.3. The function value for a new test molecule is predicted by sampling the posterior probability distribution at the test points x_{test}. Using this Bayesian approach allows us to incorporate the confidence of the predictions into the regression result. In Sect. 6.7 we will see the deep connection between Gaussian processes and Deep Neural Networks (DNN).

Recently there has been interesting work exploring the use of quantum kernels in a quantum active learning (QAL) framework with small, labelled datasets, to see if quantum computers can help speed up ML training, allow better generalization, and guide new experiments or computations for optimal design of new perovskites and doped nanoparticles [10]. PCA was employed to reduce the number of features, and each molecule in the dataset was encoded using a small number (4–8) of qubits. The qubits for each molecule were then entangled in a quantum superposition state:

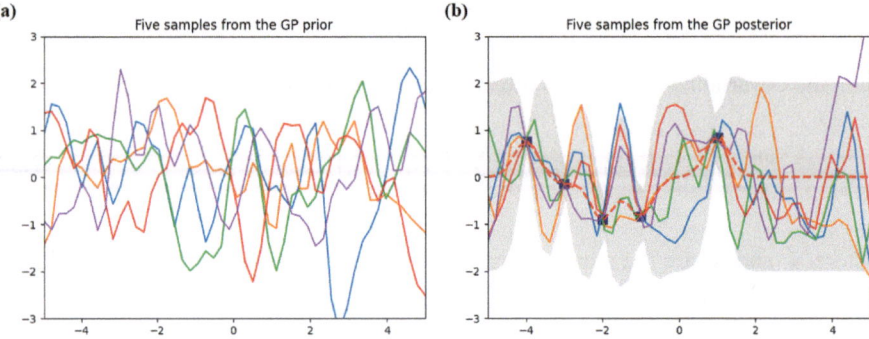

Fig. 2.3 (**a**) 5 samples from the prior of a Gaussian process regression, and (**b**) 5 samples from the posterior, assuming noiseless training data, showing the mean output (dotted line) and uncertainty (shaded grey area—2 standard deviations from the mean), for functions evaluated at 50 points, with RBF kernel, and assuming zero mean (meaning no noise added to the data). Generated by the author using code adapted from Ref. [9]

$$|\psi(x)\rangle = U(x)|0\rangle, \tag{2.60}$$

where $U(x)$ is a unitary encoding operator implemented by a quantum circuit, that rotates the initial n-qubit quantum state $|0\rangle \otimes N$ in the high-dimensional Hilbert space. Two different quantum kernels were explored, with the kernel matrix computed by measuring the inner product (2.46) of each pair of molecules on a quantum computer. This kernel matrix was then passed on to an SVR or GPR optimizer. Entanglement allows quantum feature maps and quantum kernels to represent complex patterns and potentially capture correlations between molecules that classical feature maps might miss. But the disadvantage of quantum kernels is that they are subject to noise, while classical kernels are exactly computable. In noise-free mode on a quantum emulator, QAL showed improved model performance in most, but not all, cases.

References

1. Marcus Aurelius Quotes, BrainyQuote. https://www.brainyquote.com/authors/marcus-aurelius-quotes
2. G. Maggiora, On outliers and activity cliffs: why QSAR often disappoints. J. Chem. Inf. Model. **46**(4), 1535 (2006). https://doi.org/10.1021/ci060117s
3. D. Livingstone, *Data Analysis for Chemists* (Oxford University, Press, 1995)
4. S. Wold, M. Sjöström, L. Eriksson, PLS-regression: a basic tool of chemometrics. Chemometrics Intel. Lab. Sys. **58**(2), 109–130 (2001)
5. V. Vapnik, *The Nature of Statistical Learning Theory* (Springer, New York, 1995)
6. N. Sukumar, H. Anandaram, P. Bhadra, *Computational Drug Discovery—A Primer* (Ion Cures Press, 2023)
7. C.E. Rasmussen, C.K.I. Williams, *Gaussian Processes for Machine Learning* (MIT Press, 2006). ISBN 0-262-18253-X

8. Görtler et al., A Visual Exploration of Gaussian Processes Distill (2019). https://distill.pub/2019/visual-exploration-gaussian-processes/
9. K. Bailey, Gaussian Processes for Dummies (2019). http://katbailey.github.io/post/gaussian-processes-for-dummies/
10. M.P. Lourenço, H. Zadeh-Haghighi, J. Hostaš, M. Naseri, D. Gaur, C. Simon, D.R. Salahub, Exploring Quantum Active Learning for Materials Design and Discovery. arXiv:2407.18731v1

Chapter 3
Network Measures and Spectral Graph Theory

> *Networks are present everywhere. All we need is an eye for them.*
> —*Albert-Laszlo Barabási.*

3.1 Network Measures

Networks are characterized by node attributes, edge attributes and global network measures [1, 2]. Unless otherwise stated, we will deal with only undirected, unweighted graphs, with no self-loops (*simple graphs*) in most of this chapter. The degree k_i of a node i is an attribute that we have already encountered in Sect. 1.2 (Eq. 1.3). We denote the number of nodes or vertices in a graph G by n, and the number of edges or connections between the vertices by m. An *empty graph* is one with zero edges, m = 0.

$$m = \frac{1}{2} \sum_{i=1}^{n} k_i = \frac{1}{2} \sum_{i,j=1}^{n} A_{ij}, \qquad (3.1)$$

where A_{ij} is the ij^{th} element of the adjacency matrix, and the factor ½ arises from the fact that each edge in an undirected graph is counted twice in the summations. We recall from Eqs. (1.1) and (1.2) of Sect. 1.2 that $A_{ij} = A_{ji}$ and $A_{ii} = 0$ for an undirected, simple graph. Perhaps the simplest global network measure is the *average degree*:

$$\bar{k} = \frac{\sum_{i=1}^{n} k_i}{n} = \frac{2m}{n}. \qquad (3.2)$$

The probability that any given node in a network is of degree k is given by the *degree distribution* $P(k)$. For a directed network, we need to distinguish between the in-degree and the out-degree of any node, and Eqs. (3.1) and (3.2) are replaced by:

N. Sukumar, *Navigating Molecular Networks*,
SpringerBriefs in Materials, https://doi.org/10.1007/978-3-031-76290-1_3

$$m = \sum_{i=1}^{n} k_i^{in} = \sum_{i=1}^{n} k_i^{out} = \sum_{i,j=1}^{n} A_{ij} \tag{3.3}$$

and:

$$\bar{k}^{in} = \bar{k}^{out} = \frac{m}{n}. \tag{3.4}$$

The *edge density* $\rho(G)$ of a graph G is defined as the ratio of the number of edges m to the maximum number of possible edges. Since the maximum possible number of edges in a graph is the number of pairs of nodes, which is $^nC_2 = n(n-1)/2$, the edge density is:

$$\rho(G) = \frac{2m}{n(n-1)}. \tag{3.5}$$

The edge density is a very useful measure to compare networks of different sizes. The decrease in the number of edges with the similarity cutoff (or increase with the distance threshold) has been noted earlier in Sect. 1.5. Figure 3.1 shows how the edge density varies with the distance threshold for a CSN of 974 potential Glutaminase C enzyme inhibitors [3]; this behavior is rather typical.

A *complete graph* is one where every pair of nodes is directly connected by an edge, i.e. $m = {}^nC_2$ or $\rho(G) = 1$. The *path length* $d_{i,j}$ between a given pair of nodes i and j is defined as the shortest path between them along the edges of the graph. The longest path length to be found in a graph is known as its *diameter*. Averaging the

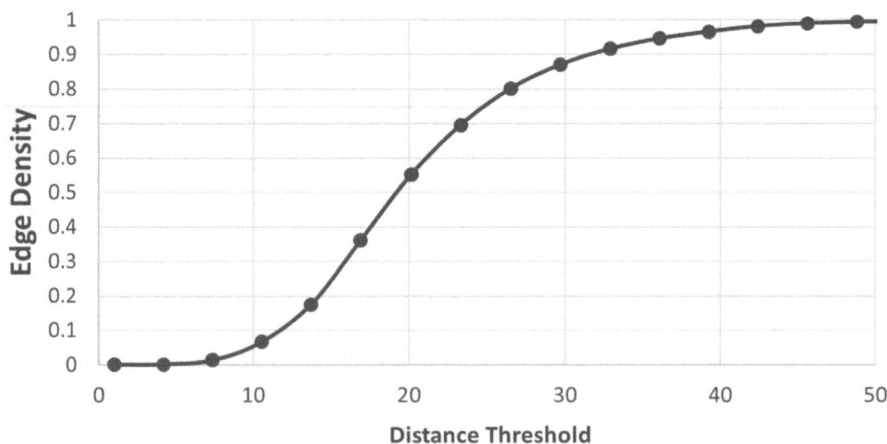

Fig. 3.1 Variation of the edge density with the distance threshold (not normalized) for a chemical dataset of 974 potential inhibitors of the Glutaminase C enzyme

path lengths $d_{i,j}$ over all n_p pairs of nodes i, j gives the *average path length* $l(G)$ of the graph G:

$$l(G) = \sum_{i,j} \frac{d_{i,j}}{n_p}. \tag{3.6}$$

An undirected graph G is known as a *connected graph* if there is a path between every pair of nodes in G; it is disconnected if there exist two nodes with no path between them. A *subgraph* of a graph G is a smaller graph formed from a subset of the nodes and edges of G. A *component* of a graph is a connected subgraph that is not contained in any larger connected subgraph. The components of a graph separate its nodes into disjoint sets. The number of components in a graph is an important graph invariant. A connected graph has only one component, which is identical to the entire graph, whereas in an empty graph, every node is a separate component with zero edges. A component of a graph G that is much larger than all other components in G is called the *giant component*.

Another node attribute is the *clustering coefficient* C_{iG}, the probability that the neighbors of a node i are directly connected to each other through an edge. It is given by the number of triangles t_{ri} that pass through the node i divided by the total number of possible triangles through i:

$$C_{iG} \frac{2 * tr_i}{(n-2)(n-1)}. \tag{3.7}$$

The *average clustering coefficient* $C(G)$ of a graph G is a global network measure that is the average of C_{iG} over all n nodes of the graph:

$$C(G) = \sum_{i=1}^{n} \frac{C_{iG}}{n} \tag{3.8}$$

An alternate measure of clustering is the global clustering coefficient or *transitivity*, given by the ratio of thrice the number of triangles in the graph to the number of all connected triples, or the fraction of all triples that are closed. Table 1.4 lists the average degree, edge density, graph diameter and transitivity values for all the networks of Fig. 1.6. It is seen that the networks in Fig. 1.6c, g have transitivity = 1, since each component is fully connected. Formally the graph diameter should be infinite for a disconnected graph (i.e. for a graph comprising two or more disconnected components). But if we consider only the largest connected component (the giant component) of a typical large CSN, the diameter of this component rapidly increases to a maximum as the edge density increases and more outlying nodes get connected to this component, and then decreases as more triangles are added, resulting in shorter paths to already connected nodes. The transitivity correspondingly shows a minimum, as shown in Fig. 3.2 for a CSN of 974 potential Glutaminase

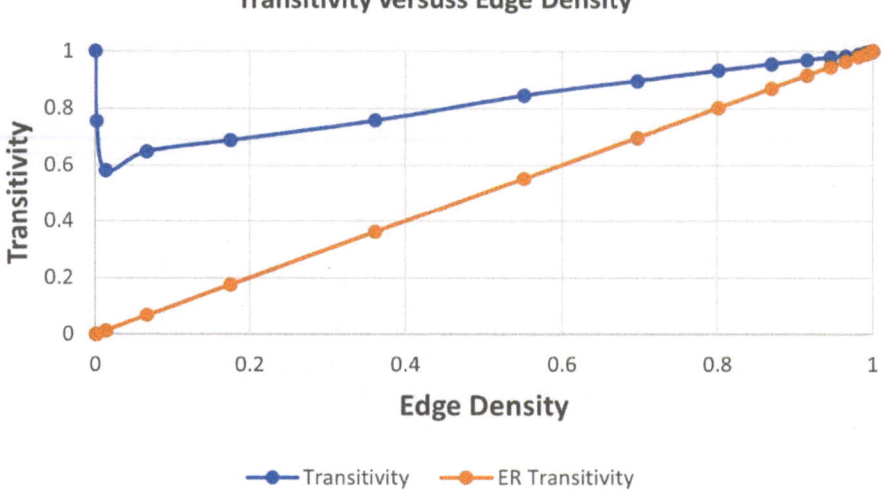

Fig. 3.2 Variation of the transitivity with the edge density for a chemical dataset of 974 potential inhibitors of the Glutaminase C enzyme

C inhibitors [3]. This maximum diameter and minimum transitivity for CSNs corresponds to the emergence of a giant component. At a very tight similarity or very short distance threshold, most of the nodes are disconnected, but as more edges are added, the network grows outwards, increasing its diameter. Eventually, with the emergence of the giant component, all or most of these components get connected, and the graph stops growing outwards. Increasing the threshold further adds triangles instead, leading to increase in the transitivity and decrease in the diameter.

Degree *assortativity* is another global network measure, given by the correlation coefficient between the degrees of pairs of nodes that are directly connected through an edge. A network is said to be *assortative* if this correlation coefficient is positive, i.e. if high degree nodes prefer connecting to one another and low degree nodes likewise prefer connecting to other low degree nodes. A *dissortative* network is one with negative assortativity, where high degree nodes preferentially connect to low degree nodes in the network. Typical CSNs tend to be assortative at low to moderate edge densities, but a dissimilarity network constructed on the same chemical space by connecting only dissimilar pairs of molecules (with similarity <u>less</u> than a specified cutoff) will show dissortative behavior [4], as seen in Fig. 3.3. All the networks Fig. 1. 6a–e are seen in Table 1.4 to be assortative, except for Fig. 1.6f, which is dissortative. This switchover from assortative to dissortative behavior with variation of similarity threshold has also been observed in larger chemical spaces [5].

The tendency of similar nodes in a network to cluster together is known as *homophily*. *Community structure* describes the existence of densely connected clusters of nodes or communities in a network, with a large number of connections between the nodes within a community, but much fewer connections between

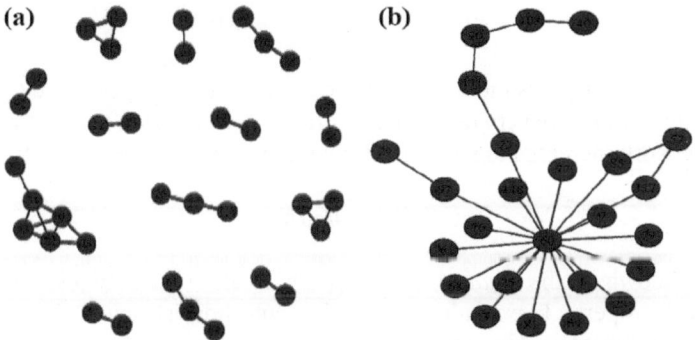

Fig. 3.3 Similarity and dissimilarity networks in chemical space. Network model of a diversity-oriented library consisting of 118 molecules, generated with MACCS fingerprints. (**a**) Similarity Network at tc $\geq$ 0.9 showing assortative behavior, homophily in a sparse network. (**b**) Dissimilarity network at tc $\leq$ 0.2 showing dissortative behavior

different communities. A useful decomposition of a network into communities is thus one that has fewer edges between communities than would be expected on the basis of random chance, and more edges within communities. *Modularity* Q is a measure of the community structure of a network:

$$Q = \frac{1}{2m} \sum_{1 \leq i,j \leq n} \left(A_{ij} - \frac{k_i k_j}{2m} \right) \delta(i,j), \tag{3.9}$$

where $\delta(i,j) = 1$ if nodes i and j belong to the same community, and $\delta(i,j) = 0$ otherwise. Here, each non-zero A_{ij} signifies the presence of an edge between the nodes i and j, while $k_i k_j/2m$ counts the expected number of edges if they were placed at random. Thus the modularity is given by the sum of the excess number of edges between all pairs of nodes within the same community, normalized by the total number of edges. It ranges between -1 for graphs with very low community structure and 1 for graphs with very high community structure. Community structure detection reveals the structure of a large network that can be naturally partitioned into communities. Detection and characterization of modularity in networks is of great practical significance. Community structure in a metabolic or gene regulatory network can provide evidence for modularity in the dynamics of the network [6–8], with different groups of nodes (communities) acting relatively independently to perform different metabolic or regulatory functions.

Different global network measures characterize different classes of networks. Erdös and Rényi [9–11] studied a random network with n nodes connected by m edges that are chosen randomly with a probability p from the $^{n}C_2$ possible edges. Alternatively, for each pair of nodes, we can assign randomly and independently, either an edge with probability p or no edge with probability $1 - p$. The connectivity of an *Erdős–Rényi random network* depends on p, and the degree distribution follows *Poisson statistics*, falling off asymptotically with the degree k as:

$$P(k) \sim e^{-k} \text{ as } k \to \infty, \tag{3.10}$$

indicating that the degrees most of nodes are close to the average degree $\bar{k}$, deviating significantly from $\bar{k}$ only very rarely. The average path length $l(G)$ of the Erdös-Rényi random network increases logarithmically with the size n of the network:

$$l(G) \sim \log(n). \tag{3.11}$$

This is known as the *small-world property*. Real-world networks also display small-world property like random networks, but show stronger clustering, with clustering coefficients independent of n.

Watts and Strogatz [12] introduced a sparse, small-world network, with high clustering coefficients and small diameter by starting with n nodes on a ring with only local edges, *i.e.* where each node is connected to its nearest z neighbors on either side ($z \ll \underline{n}$), and then randomly rewiring each edge with a probability p, avoiding self-loops and duplicate edges. This introduces a few (p * n * z) long-range edges into the hitherto local network. As the probability p is varied, the network transitions from ordered (p = 0) to random (p = 1). Giant components are frequently found in random networks, as well as in real-world networks. Random networks are also characterized by the existence of a percolation threshold—a value of the probability p above which a giant component emerges and below which it does not.

Scale-free networks [13] have degree distributions that decay as a *power law* with the degree k:

$$P(k) \sim k^{-\gamma} \text{ as } k \to \infty, \tag{3.12}$$

where γ is a power law exponent. Such a degree distribution appears linear on a log–log plot, as in Fig. 3.4, which shows log $P(k)$ versus log k for over 15 million molecules of the ZINC database [5].

The number of highly connected nodes in such networks is much greater than for a random graph, and therefore such a distribution is often called a fat-tailed distribution. The dynamics of scale-free networks are dominated by a small number of high-degree nodes, or *hubs*. Such networks can form by a process of *preferential attachment* [14]; when new nodes are added to the network, they attach preferentially to nodes of high degree. The clustering coefficients C_{iG} of a scale-free network are inversely proportional to the node degree k_i. Real-world networks probably arise from a competition between homophily and preferential attachment [15]. The average path lengths of scale-free networks are even shorter than for Erdös-Rényi random networks:

$$l(G) \sim \log(\log(n)). \tag{3.13}$$

CSNs are seen to display the small-world property, as well as power law degree distributions [5, 16, 17].

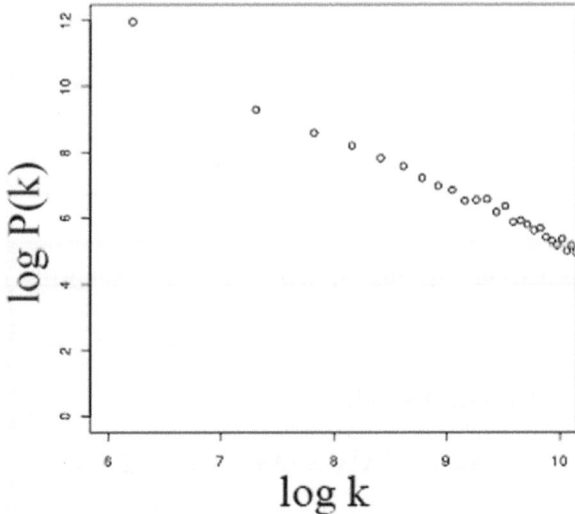

A class of networks with local clustering, high modularity and scale-free topology is the *hierarchical network*. In such networks, the clustering coefficient scales as:

$$C(k) \sim k^{-1} \tag{3.14}$$

appearing as a straight line on a log–log plot, with slope -1. Communication between different clusters in a hierarchical network is mediated by the hubs of the network.

3.1.1 Eigenvalues of the Adjacency Matrix

The *graph spectrum* is the set of eigenvalues of the adjacency matrix **A** associated with a graph. The density of these eigenvalues is referred to as the graph spectral density. Eigenvalues encode information about the fundamental topological properties of the associated graph. Since an undirected graph has a real, symmetric adjacency matrix **A** (Eq. 1.2), its eigenvalues are real. We can establish this result by first noting that since **A** is real and symmetric,

$$\mathbf{A} = \mathbf{A}^{T} = \mathbf{A}^{*}. \tag{3.15}$$

If **x** is a non-trivial (i.e. non-zero) eigenvector of **A** with eigenvalue λ, then we can write:

$$\mathbf{A}\mathbf{x} = \lambda\mathbf{x}, \tag{3.16}$$

whence,

$$(\mathbf{Ax})^* = (\lambda\mathbf{x})^* \tag{3.17}$$

or, using (3.15),

$$\mathbf{Ax}^* = \lambda^*\mathbf{x}^*. \tag{3.18}$$

Left multiplying Eq. (3.16) by $\mathbf{x}^{*T}$, we can write:

$$\mathbf{x}^{*T}\mathbf{Ax} = \mathbf{x}^{*T}\lambda\mathbf{x} = \lambda\left(\mathbf{x}^{*T}.\mathbf{x}\right) = \lambda||\mathbf{x}||^2. \tag{3.19}$$

But since $\mathbf{A} = \mathbf{A}^T$,

$$\mathbf{x}^{*T}\mathbf{Ax} = \mathbf{x}^{*T}\mathbf{A}^T\mathbf{x} = (\mathbf{Ax}^*)^T\mathbf{x} = (\lambda^*\mathbf{x}^*)^T\mathbf{x} = \lambda^*(\mathbf{x}^{*T}.\mathbf{x}) = \lambda^*||\mathbf{x}||^2, \tag{3.20}$$

where we have utilized Eq. (3.18). Subtracting Eq. (3.20) from (3.19) gives:

$$(\lambda - \lambda^*)||\mathbf{x}||^2 = 0. \tag{3.21}$$

This is possible for non-zero eigenvectors $\mathbf{x}$ only if:

$$\lambda = \lambda^*. \tag{3.22}$$

Thus the eigenvalues of the adjacency matrix for an undirected network are real. Another important result we should note is that the eigenvalues are invariant to permutations of the order of the atoms. For a directed network, however, the adjacency matrix is not symmetric, and its eigenvalues can be real or complex.

We consider here adjacency matrices $\mathbf{A}$ of undirected graphs of size n, and eigenvalues and eigenvectors of the secular equation:

$$\det.||\mathbf{A} - \lambda\mathbf{I}|| = 0, \tag{3.23}$$

where $\mathbf{I}$ is the n × n identity matrix. The *trace*, or the sum of all the eigenvalues of the adjacency matrix $\mathbf{A}$, including multiplicities, is always 0:

$$\text{Tr}(\mathbf{A}) = \sum_{i=1}^{n} A_{ii} = \sum_{i=1}^{n} \lambda_i = 0. \tag{3.24}$$

The product of all the eigenvalues, again including multiplicities, is the determinant of $\mathbf{A}$:

$$\prod_{i=1}^{n} A_{ii} = \det.||\mathbf{A}||. \tag{3.25}$$

The number of non-zero eigenvalues, again counting all multiplicities, is known as the *rank* of the matrix. It is easily shown that if the diameter of a graph is d, then the number of distinct eigenvalues of $\mathbf{A}$ has to be at least $d + 1$. The eigenvalue spectrum of a bipartite graph is symmetric about 0. For a connected graph G, the largest eigenvalue λ_{max} of its adjacency matrix $\mathbf{A}$ is bounded by the maximum and average node degrees:

$$\bar{k} \le \lambda_{max} \le k_{max}. \tag{3.26}$$

The *complement* of a graph G is another graph G' defined on the same set of nodes, such that there is an edge between a pair of nodes in G' iff there is no edge between them in G. The graph complement of the complete graph K_n with n nodes is the empty graph on n nodes (i.e. with no edges). The union of a graph G and its complement G' is a complete graph, whereas the intersection of two complement graphs is the empty graph. The complement of any disconnected graph is connected. The size of a graph and its complement are the same and equal to the number of nodes. A complete bipartite graph $K_{m,n}$ has a rank 2 adjacency matrix, and therefore an eigenvalue 0 of multiplicity $n - 2$, and two non-trivial eigenvalues $\pm \lambda$, since the eigenvalues have to sum to 0, as per Eq. (3.24).

A *regular* graph is one where every node has the same degree: $k_i = k$ for all i. Equation (1.3) then implies that $\mathbf{A}e = k\mathbf{e}$, where $\mathbf{e}$ is a vector all of whose elements are 1's:

$$\mathbf{e} = (1, 1, 1, \ldots 1)^{\mathrm{T}}. \tag{3.27}$$

Thus $\mathbf{e}$ is an eigenvector of the adjacency matrix $\mathbf{A}$ for a regular graph, with eigenvalue $\lambda_{max} = k_{max} = k$. If G is regular of degree k and size n, then G' is regular of degree $n - 1 - k$, and:

$$A(G') = \mathbf{J} - \mathbf{I} - A(G), \tag{3.28}$$

where $\mathbf{I}$ is, as before, the $n \times n$ identity matrix and $\mathbf{J}$ is the $n \times n$ matrix all of whose elements are 1's. All other eigenvectors are orthogonal to $\mathbf{e}$, and hence if the eigenvalues of G are $\lambda_1, \lambda_2, \ldots, \lambda_{max} = k$, then the eigenvalues of its complement graph G' are $n - 1 - k, -1 - \lambda_{n-1}, \ldots, -1 - \lambda_2, -1 - \lambda_1$.

A subgraph is a component of a graph unconnected to any node not contained in the subgraph. A subgraph with n nodes and m edges will be denoted g(n, m). The following results can easily be verified by expansion of the secular determinant:

1. The eigenvalues of a graph composed of disconnected components (subgraphs) are the eigenvalues of the individual components (subgraphs).
2. Each unconnected node (of degree 0) will contribute an eigenvalue 0.
3. Each isolated edge, i.e. subgraph g(2,1), will contribute a pair of eigenvalues $\pm$ 1.

Table 3.1 Algebraic connectivity, edge connectivity and vertex connectivity for some special types of graphs

Graph	Example	a(G)	e(G)	v(G)
Linear chain		$2(1 - \cos(\pi/n))$	1	1
Ring		$2(1 - \cos(2\pi/n))$	2	2
Star		1	1	1
Complete graph		n	$n - 1$	$n - 1$

4. Each completely connected subgraph, g(n,m) with $m = \frac{1}{2} n(n - 1)$, contributes an eigenvalue $- 1$ of degeneracy $n - 1$ and an eigenvalue $n - 1$.

5. Each star subgraph (Table 3.1), i.e. g(n, n − 1) with one node of degree $n - 1$ connected to all the other n − 1 nodes, each of degree 1, of the subgraph, contributes an eigenvalue 0 of degeneracy n-2 and a pair of eigenvalues $\pm \sqrt{(n - 1)}$.

6. Each linear chain subgraph (Table 3.1), i.e. g(n, n − 1) with 2 nodes of degree 1 and $n - 2$ nodes of degree 2, contributes eigenvalues $- 2\cos j\pi/(n + 1)$, with j $= 1, 2, \ldots$ n.

7. Each cyclic subgraph (Table 3.1), i.e. g(n,n) with all nodes of degree 2, contributes eigenvalues $2\cos 2j\pi/n$, with j $= 0, 1, 2, \ldots$ n/2, and all eigenvalues except j $= 0$ being doubly degenerate.

3.1.2 Eigenvalues of the Laplacian Matrix

The use of eigenvectors of the Laplacian matrix $\mathbf{L} = \mathbf{D} - \mathbf{A}$ (Eq. 1.4) to probe the structure of a graph was pioneered by Fiedler [18]. Consider the vector $\mathbf{e}$ of Eq. (3.27), all of whose elements are ones. As pointed out in Sect. 1.3, the row sum of the Laplacian matrix is zero. So we have:

$$\mathbf{L}\mathbf{e} = \mathbf{D}\mathbf{e} - \mathbf{A}\mathbf{e} = 0, \tag{3.29}$$

implying that:

$$\lambda_1 = 0 \tag{3.30}$$

is an eigenvalue of $\mathbf{L}$ corresponding to the eigenvector $\mathbf{e}$. From Eq. (1.15), we have, for any non-zero vector $\mathbf{x}$:

$$\mathbf{x}^T \mathbf{L} \mathbf{x} = \mathbf{x}^T \mathbf{N} \mathbf{N}^T \mathbf{x} = \left(\mathbf{N}^T \mathbf{x}\right)^T \left(\mathbf{N}^T \mathbf{x}\right) = \left\|\mathbf{N}^T \mathbf{x}\right\|^2 \geq 0, \tag{3.31}$$

which means that the Laplacian matrix $\mathbf{L}$ is positive semi-definite. Since $\mathbf{L}$ is symmetric, there exists an orthonormal basis $\mathbf{x_1}, \mathbf{x_2}, \ldots \mathbf{x_n}$ of eigenvectors of $\mathbf{L}$ such that:

$$\mathbf{x_i^T x_j} = 0 \qquad \text{for } i \neq j,$$
$$\text{and} \qquad \mathbf{x_i^T x_i} = 1. \tag{3.32}$$

with the corresponding eigenvalues $\lambda_1, \lambda_2, \lambda_3, \ldots \lambda_n$:

$$\mathbf{L} \mathbf{x_i} = \lambda_i \mathbf{x_i}. \tag{3.33}$$

Since $\mathbf{L}$ is positive semi-definite, we have, from Eqs. (3.31), (3.32) and (3.33):

$$\mathbf{x_i^T L x_i} = \mathbf{x_i^T} (\lambda_i \mathbf{x_i}) = \lambda_i \mathbf{x_i^T}.\mathbf{x_i} = \lambda_i \geq 0. \tag{3.34}$$

Thus all eigenvalues of $\mathbf{L}$ are non-negative. So the eigenvalues λ_i of $\mathbf{L}$ can be ordered as follows:

$$0 = \lambda_1 \leq \lambda_2 \leq \lambda_3 \leq \ldots \leq \lambda_n = \lambda_{max} \tag{3.35}$$

A graph G is connected iff $\lambda_1 = 0$ is a simple or non-degenerate eigenvalue of $\mathbf{L}$. In other words, the Laplacian matrix of a connected graph has one and only one zero eigenvalue. The multiplicity or degeneracy of the zero eigenvalue λ_1 of $\mathbf{L}$ gives the number of connected components in the graph G. This can be easily shown as follows:

$$\mathbf{L} \mathbf{x_i} = k_i \mathbf{x_i} - \sum_{j:(i,j)\in E} \mathbf{x_j} = \sum_{j:(i,j)\in E} \left(\mathbf{x_i} - \mathbf{x_j}\right). \tag{3.36}$$

where the sums are over all edges incident on node i, whence:

$$\mathbf{x}^T \mathbf{L} \mathbf{x} = \sum_i \mathbf{x_i^T} . \sum_{j:(i,j)\in E} \left(\mathbf{x_i} - \mathbf{x_j}\right)$$
$$= \sum_{(i,j)\in E} \mathbf{x_i^T} . \left(\mathbf{x_i} - \mathbf{x_j}\right)$$

$$= \sum_{i<j:(i,j)\in E} \left\{ \mathbf{x}_i^T \cdot (\mathbf{x}_i - \mathbf{x}_j) + \mathbf{x}_j^T \cdot (\mathbf{x}_j - \mathbf{x}_i) \right\}$$

$$= \sum_{i<j:(i,j)\in E} \left\{ \mathbf{x}_i^T \cdot (\mathbf{x}_i - \mathbf{x}_j) - \mathbf{x}_j^T \cdot (\mathbf{x}_i - \mathbf{x}_j) \right\}$$

$$= \sum_{i<j:(i,j)\in E} (\mathbf{x}_i - \mathbf{x}_j)^2. \tag{3.37}$$

This expression can be zero only if $\sum_{(i,j)\in E} (\mathbf{x}_i - \mathbf{x}_j)^2 = 0$ separately for each component of G. Hence, $\mathbf{x}_i = \mathbf{x}_j$ for all eigenvectors with zero eigenvalue, in each component. So if G is connected, then all $\mathbf{x}$ in G with zero eigenvalue are equal. Now, as shown in Eq. (3.29), $\mathbf{e}$ is an eigenvector of $\mathbf{L}$ with eigenvalue 0; so $\mathbf{e}^T \mathbf{L} \mathbf{e} = 0$. Thus all $\mathbf{x}$ in G with zero eigenvalue are multiples of $\mathbf{e}$, implying that the multiplicity of the zero eigenvalue λ_1 is 1, *i.e.* that λ_1 is a simple eigenvalue. Conversely, if the graph G consists of q disconnected components $G_1, G_2, G_3, \ldots G_q$, then let us define $\mathbf{e}_i$ as the vector with entries equal to 1 on each node of component G_i and equal to 0 otherwise. Then $\mathbf{N}^T \mathbf{e}_i = 0$ and thus $\mathbf{L} \mathbf{e}_i = \mathbf{N} \mathbf{N}^T \mathbf{e}_i = 0$. Since $\mathbf{e}_1, \mathbf{e}_2, \ldots \mathbf{e}_q$ are linearly independent, the multiplicity of the zero eigenvalue λ_1 is at least q. However, since each component G_i is connected, and we have just shown that for a connected graph λ_1 is a simple eigenvalue, therefore the multiplicity of λ_1 is exactly q.

The second-smallest eigenvalue of the Laplacian matrix of G, with multiple eigenvalues counted separately, is called the *algebraic connectivity* $a(G) = \lambda_2$ [18] or the *Fiedler eigenvalue* of the graph. $a(G) > 0$ iff G is a connected graph. In other words, $\lambda_2 = 0$ iff G is disconnected. This is because if G is disconnected, we can divide it into disconnected components G_1 and G_2 consisting of n_1 and n_2 nodes respectively, with no edges between them. If we now re-index the nodes so as to block out the matrix $\mathbf{L}$ into two blocks of sizes $n_1 \times n_1$ and $n_2 \times n_2$ with no matrix elements between them, then the vectors $\mathbf{e}_1 = (1, 1, 1, \ldots 0, 0, 0)$ with n_1 ones followed by n_2 zeroes and $\mathbf{e}_2 = (0, 0, 0, \ldots 1, 1, 1)$ with n_1 zeroes followed by n_2 ones are both eigenvectors of $\mathbf{L}$ associated with the zero eigenvalue: $\mathbf{L} \mathbf{e}_1 = \mathbf{L} \mathbf{e}_2 = 0$, and orthogonal to each other. Thus $\lambda_2 = 0$ when G is a disconnected graph.

Using similar arguments, it can be shown that $\lambda_q = 0$ iff G has at least q components. *Edge connectivity* $e(G)$ is defined as the minimum number of edges whose deletion results in loss of connectivity of the graph G. Likewise, the *vertex connectivity* $v(G)$ of G is the minimum number of nodes (vertices) whose deletion, together with their adjacent edges, causes fragmentation of the graph into disconnected components. For the complete graph K_n, all of whose n nodes are directly connected by an edge, $v(G) = n - 1$ and $a(K_n) = n$. For a non-complete graph G, the algebraic connectivity, edge and vertex connectivities are related through the inequality:

$$a(G) \leq v(G) \leq e(G) \tag{3.38}$$

The eigenvalues of the Laplacian are used to derive bounds on various types of cuts in graphs. If G' is the complement of G and $\mathbf{L}'$ is the Laplacian matrix of G', then it is easy to show that:

$$L + L' = nI - J, \tag{3.39}$$

where, as before, I is the $n \times n$ identity matrix and J is the $n \times n$ matrix all of whose elements are 1's. Therefore:

$$\lambda_{max} \leq n. \tag{3.40}$$

The Laplacian spectrum of the complement graph G' is then:

$$\lambda_i(G') = n - \lambda_{n-i}(G) \text{ for } i = 1, \cdots, n - 1. \tag{3.41}$$

Another upper bound to λ_{max} for a simple graph is provided by the maximum of the sum of the degrees of any pair (i, j) of connected nodes in the graph:

$$\lambda_{max} \leq \max_{(i, j) \in E(G)} (k_i + k_j). \tag{3.42}$$

Yet another upper bound to λ_{max} is provided by the maximum of the sum of the degree of a node and the average k_i'' of the degrees of all its adjacent nodes:

$$\lambda_{max} \leq \max (k_i + k_i''). \tag{3.43}$$

The algebraic connectivity satisfies several bounds:

$$a(G) \geq 2 \min k_i(G) - n + 2, \tag{3.44}$$

$$a(G) \leq \left(\frac{n}{n-1}\right) \min k_i \leq \frac{2m}{n-1}, \tag{3.45}$$

where k_i is the degree of the ith node, and m is the number of edges in G. Also:

$$(\frac{n}{n-1}) \max k_i \leq \lambda_{max} \leq \max k_i. \tag{3.46}$$

The maximum eigenvalue λ_{max} of L of G is related to the algebraic connectivity of the complement G' of G:

$$\lambda_{max} = n - a(G'). \tag{3.47}$$

Also, since $a(G) = \lambda_2$:

$$a(G) \leq \lambda_{max}, \tag{3.48}$$

with the equality holding iff G is a complete graph or an empty graph. If G_i are all components of G with the same set of nodes,

$$\lambda_{max} = \max_i \lambda_{max}(G_i), \tag{3.49}$$

$$\lambda_{max}(G_1) \leq \lambda_{max}(G_2) \text{ if } G1 \subseteq G2, \tag{3.50}$$

$$\lambda_{max}(G_1 \cup G_2) \leq \lambda_{max}(G_1) + \lambda_{max}(G_2) - a(G_1 \cap G_2). \tag{3.51}$$

If G_1 is a subgraph of G on the same set of nodes, and if $\mu_1 \leq \mu_2 \leq \cdots \leq \mu_n$ are the eigenvalues of the Laplacian of G_1, then $\lambda_i \geq \mu_i$ for all $i \leq n$. If G_1 and G_2 are graphs on the same set of nodes that do not share an edge, then:

$$a(G_1) + a(G_2) \leq a(G_1 \cup G_2). \tag{3.52}$$

The algebraic connectivity a(G) is a non-decreasing function for graphs on the same set of nodes, i.e. $a(G_1) \leq a(G_2)$ if $G_1 \subseteq G_2$ and G_1, G_2 have the same set of nodes. If G_1 arises from a graph G by removing n' vertices from G along with all their adjacent edges, then:

$$a(G_1) \geq a(G) - n'. \tag{3.53}$$

If the graph G with n nodes contains a set of n'' nodes such that no two of them are joined by an edge of G, then:

$$a(G) \leq n - n''. \tag{3.54}$$

It therefore follows that for non-complete a graph G with n nodes,

$$a(G) \leq n - 2, \tag{3.55}$$

since there is at least one pair of nodes that is not connected by an edge.

For a complete bipartite graph K_{qs} with q nodes of one kind fully connected to s nodes of another kind:

$$a(K_{qs}) = \min(q, s). \tag{3.56}$$

Table 3.1 lists the values of the algebraic connectivity, edge connectivity and vertex connectivities for some special types of graphs.

3.2   Graph Centrality Measures

Network centrality measures are node attributes that are useful for locating the most important nodes in a network, and also for identifying functional clusters in complex networks. Many different centrality measures have been proposed [1], such as the vertex degree centrality and centralization, betweenness centrality, closeness centrality, subgraph centrality and eigenvector centrality. The *degree centrality* is a count of the number of links that a node has within the network:

$$k_i = \sum_{j=1}^{n} A_{ij}. \tag{3.57}$$

It can be interpreted as a measure of immediate influence of a node on a network. Disabling a critical fraction of the nodes in a random network results in a phase transition and functional disintegration of the network, breaking it up into isolated components that are no longer in communication with each other. Scale-free networks, on the other hand, are robust against random or accidental failures, which mainly affect the many low degree nodes [19]. Even if 80% of randomly selected nodes fail, the network's integrity is not disrupted, as the remaining nodes still form a connected cluster, and such systems are resilient against random component failure. But this renders scale-free networks, such as the internet, vulnerable to targeted attack, such as denial of service attacks targeting a few key hubs with high k_i, and such targeted attacks can disrupt communication by breaking up the network into small, isolated components.

Degree Centralization is the average deviation of the node degrees from the maximum degree, normalized to its maximum value ($n^2 - 3n + 2$) for a star graph (a graph with all nodes connected to a central node of degree $n - 1$, as shown in Table 3.1). Figure 3.5 shows the variation of the degree centralization with the edge density for a chemical dataset of 974 potential Glutaminase C inhibitors. While the degree centralization of Erdős–Rényi networks is generally very small, those for CSNs increase gradually to a maximum before gradually decreasing again, indicating that CSNs are much more "starry" or consist of more hubs (or hubs of higher degree) than Erdős–Rényi networks.

Closeness centrality Cc of a node i, a measure applicable to connected networks, is the reciprocal of the sum of the shortest path lengths $d_G(i, j)$ from i to every other node j, and describes the closeness of the node i to all other nodes in the graph:

$$Cc(i) = \frac{1}{\sum_{j \neq i} d_G(i, j)} \tag{3.58}$$

The closeness centrality can also be defined for edges of a graph, to give a measure of the closeness of an edge to all other edges in the graph.

Betweenness centrality C_B of a node i measures how often the node i occurs on the shortest path between all pairs of nodes in the graph:

$$C_B(i) = \frac{\sigma(i)}{\sigma}, \tag{3.59}$$

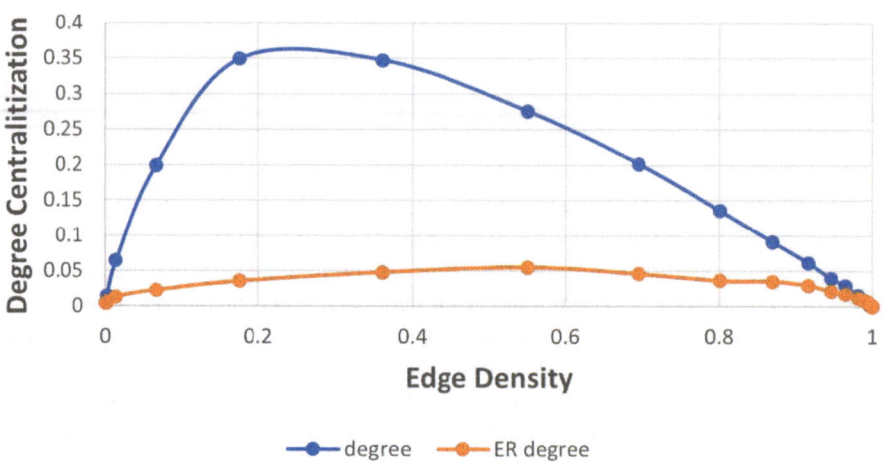

Fig. 3.5 Variation of the Degree Centralization with the edge density for a chemical dataset of 974 potential inhibitors of the Glutaminase C enzyme

where σ denotes the number of shortest paths between all pairs of nodes in the graph, and $\sigma(i)$ is the number of these paths that pass through the node i. Figure 3.6 displays the betweenness centrality of nodes for the giant component of Fig. 1.6d. It is seen that unconnected and outlying nodes have zero betweenness centrality, and nodes with high degree generally have higher betweenness centrality, but nodes with the same degree need not have the same betweenness centrality, as can be seen by inspecting the three nodes of degree 3 in Fig. 3.6. Betweenness centrality can likewise be defined for edges by counting the frequency of occurrence of an edge on the shortest path between all pair of nodes. Betweenness centrality quantifies how influential a node (or edge) is in effecting communication between other pairs of nodes (or edges). Most networks are vulnerable to targeted deletion of nodes with high degree or high betweenness centrality; deleting nodes of high degree or high betweenness centrality results in rapid disintegration of the network. In CSNs, molecules with high betweenness centrality serve as "chemical bridges" common to different QSARs [20], and are important for scaffold hopping (Sect. 5.1). In metabolic and regulatory networks, they can act as chemical bottlenecks.

Subgraph centrality [21] describes nodes based on their participation in all subgraphs of the network, assigning larger weights to smaller subgraphs, making this an appropriate measure for characterizing the various motifs in a network. The subgraph centrality of a node i is obtained by summing the number of closed paths of length q starting and ending on node i in the network, which is given by the ith diagonal element of the qth power of the adjacency matrix $\mathbf{A}$:

$$\mu_q(i) = (\mathbf{A}^q)_{ii}. \qquad (3.60)$$

Fig. 3.6 Betweenness
centrality of nodes for the
giant component of Fig. 1.6d

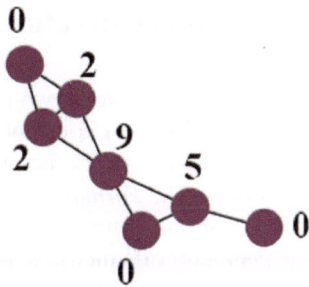

$\mu_q(i)$ is known as the local spectral moment. To avoid divergence, the contribution of each closed path is scaled by dividing it by the factorial of the order of the spectral moment (q!). Each closed path is associated with a connected subgraph; so subgraph centrality counts the number of times a node participates in the various connected subgraphs in the network, with smaller subgraphs assigned larger weightage. The subgraph centrality of node i in the network is thus:

$$SC(i) = \sum_{q=0}^{\infty} \frac{\mu_q(i)}{q!} \leq \sum_{q=0}^{\infty} \frac{\lambda^q}{q!} = e^{\lambda}, \tag{3.61}$$

and is bounded from above by $SC(i) \leq e^{\lambda}$, where λ is the main eigenvalue of $\mathbf{A}$. Subgraph centrality can also be evaluated from the spectra of the adjacency matrix $\mathbf{A}$:

$$SC(i) = \sum_{j=1}^{n} \left(v_j^i\right)^2 e^{\lambda_j}, \tag{3.62}$$

where v_j^i is the ith component of v_j, the jth orthonormal eigenvector of $\mathbf{A}$. $SC(i)$ is more discriminative than other vertex measures such as degree, betweenness, closeness or eigenvector centrality. The subgraph centrality displays a power-law distribution for real-world complex networks.

While degree centrality ranks nodes in order of their degree, *eigenvector centrality* is based on spectral analysis of the graphs, or eigen-decomposition of their matrix representations, and ranks nodes in order of the eigenvalues of the associated matrices:

$$x_i = \frac{1}{\lambda} \sum_{j=1}^{n} A_{ij} x_j, \tag{3.63}$$

or in matrix form:

$$\lambda \mathbf{x} = \mathbf{A}\mathbf{x}. \tag{3.64}$$

3.3   Graph Curvature

A connection between two points in continuum space (for example two molecules in chemical space) is established by the vector connecting the two points. When dealing with a network representation, the analogous connection is the edge between two nodes. The *Forman-Ricci curvature* [22] and *Ollivier's curvature* [23] represent discretizations of the Ricci curvature tensor (Sect. 2.1), which measures how much the geometry of a metric tensor differs locally from that of Euclidean space, or how a shape is deformed as one moves along the shortest paths (geodesics) in the space. While most other centrality measures quantify the importance of nodes in a graph, the Forman-Ricci curvature applies to edges as well as to nodes of weighted and unweighted graphs [24, 25] and hypergraphs [26]. The Forman–Ricci curvature of an edge e between two nodes v_1 and v_2 quantifies the flow along that edge in the network, weighted by the edges adjacent to e:

$$F(e) = \left\{ w_{v1} + w_{v2} - \sqrt{w_e} \sum_{e_{v1}, e_{v2}} \left(\sqrt{w_{v1}} + \sqrt{w_{v2}} \right) \right\}, \qquad (3.65)$$

where w_e is the weight associated with the edge e, w_{v1} and w_{v2} are the weights of the nodes v_1 and v_2, respectively, and the sum runs over e_{v1} and e_{v2}, the edges other than e incident on the nodes v_1 and v_2, respectively.

The Forman curvature defined in Eq. (3.65) for an edge can be extended to nodes by averaging the Forman curvatures of all the k_i edges e_i connected to the node *i*:

$$F(i) = \frac{1}{k_i} \sum_{e_i} F(e_i), \qquad (3.66)$$

where k_i is the degree of node *i*.

Figure 3.7 displays a histogram of the Forman-Ricci curvature for a weighted network of 13 molecules constructed from the Tanimoto similarity matrix in Table 1. 3a. It is seen that most of the 78 edges have zero or negative curvature. Negative Forman-Ricci curvatures were also found by Sreejith *et al.* [24, 25] for most nodes and edges in weighted and unweighted undirected networks; they found a narrow distribution of both node and edge curvatures in Erdös-Rényi networks and Watts–Strogatz small-world networks, but a broad distribution in scale-free networks and several real-world networks, including the yeast protein–protein interaction network. Forman curvature can thus be used to distinguish between networks of different kinds. The Forman curvature was negatively correlated with degree and centrality measure in most networks, but uncorrelated with clustering coefficient. All networks were found to be sensitive to targeted removal of nodes with large negative Forman curvatures.

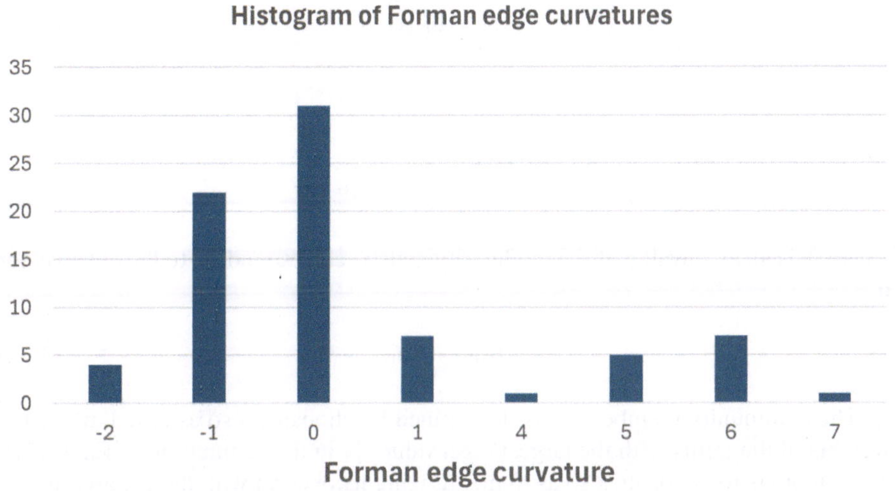

Histogram of Forman edge curvatures

Fig. 3.7 Histogram of the Forman-Ricci curvature for the edges of a weighted network of 13 molecules constructed from the Tanimoto similarity matrix in Table 1.3a

3.4 Eigenvectors of the Modularity Matrix

If we consider a network consisting of two communities, Eq. (3.9) for the Modularity can be cast into the matrix form [27]:

$$Q = \frac{1}{4m} s^T B s, \tag{3.67}$$

where s is a column vector, with elements $s_i = 1$ if the node i belongs to the first community and $s_i = -1$ if it belongs to the second community, and B is the modularity matrix with elements:

$$B_{ij} = A_{ij} - \frac{k_i k_j}{2m}. \tag{3.68}$$

Like the graph Laplacian L, the modularity matrix B too is a real symmetric matrix, with each of its rows and columns summing to zero. So the vector $e = (1, 1, 1, ...)^T$ of Eq. (3.27) is an eigenvector of B with eigenvalue zero.

Expanding s as a linear combination of the eigenvectors u_i of B:

$$s = \sum_{i=1}^{n} a_i u_i, \tag{3.69}$$

with all u_i having unit norm, and

$$a_i = \mathbf{u}_i^T.\mathbf{s}, \tag{3.70}$$

so that:

$$Q = \frac{1}{4m} \sum_i a_i \mathbf{u}_i^T \mathbf{B} \sum_j a_j \mathbf{u}_j = \frac{1}{4m} \sum_{i=1}^{n} \left(\mathbf{u}_i^T.\mathbf{s} \right)^2 \beta_i, \tag{3.71}$$

where β_i is an eigenvalue of the modularity matrix, corresponding to the eigenvector $\mathbf{u}_i$:

$$\mathbf{B}\mathbf{u}_i = \beta_i \mathbf{u}_i. \tag{3.72}$$

The community membership is determined by choosing $\mathbf{s}$ so as to maximize the weights of the terms with the largest eigenvalues β_i in the summation of Eq. (3.71). This amounts to choosing $\mathbf{s}$ so as to maximizing its overlap with the eigenvector $\mathbf{u}_1$ with the largest eigenvalue β_i, all other terms being zero due to mutual orthogonality of the eigenvectors. So we maximize the dot product $\mathbf{u}_1^T.\mathbf{s}$ subject to $s_i = \pm 1$, by setting $s_i = 1$ if $\mathbf{u}_1$ is positive and $s_i = -1$ if it is negative.

In a general network with multiple communities, the community structure is found by a recursive process. Nodes with large coefficients in the leading eigenvector (the one with most positive eigenvalue) contribute the most to modularity. Of course, the eigenvector $\mathbf{e}$ with zero eigenvalue is always a trivial solution, and all other eigenvectors will be orthogonal to it. If $\mathbf{B}$ has no positive eigenvalues, all terms in the summation of Eq. (3.71) will be zero or negative and no further division of the network can increase the modularity; the leading eigenvector is then $\mathbf{e}$, corresponding to all the nodes belonging to a single community. Such a network is said to be indivisible. So in recursive partitioning of the network, the leading (most positive) eigenvalue indicates when to stop the partitioning process: if the leading eigenvalue is zero, the network is not divisible into further communities.

This chapter has dealt with several types of network measures and graph eigenvalues. In the next chapter we will continue discussion of the eigenvalues of matrices associated with graphs, but specifically from the standpoint of Random Matrix theory (RMT).

References

1. M.E. Newman, The mathematics of networks. New Palgrave Encyclopedia Econ. **2**, 1–12 (2008)
2. N. Sukumar, M.P. Krein, G. Prabhu, S. Bhattacharya, S. Sen, Network measures for chemical library design. Drug Dev. Res. **75**, 402–411 (2014)
3. M. Kothiyal, S. Kumar, N. Sukumar, Investigation of chemical space networks using graph measures and random matrix theory. J. Math. Chem. **60**, 891–914 (2022). https://doi.org/10.1007/s10910-022-01341-y

4. G. Prabhu, S. Bhattacharya, M.P. Krein, N. Sukumar, Investigation of similarity and diversity threshold networks generated from diversity-oriented and focused chemical libraries. J. Math. Chem. **54**(10), 1916–1941 (2016). https://doi.org/10.1007/s10910-016-0657-0

5. M.P. Krein, N. Sukumar, Exploration of the topology of chemical spaces with network measures. J. Phys. Chem. A **115**, 12905 (2011). https://doi.org/10.1021/jp204022u

6. P. Holme, M. Huss, H. Jeong, Subnetwork hierarchies of biochemical pathways. Bioinformatics **19**, 532–538 (2003). https://doi.org/10.1093/bioinformatics/btg033

7. R. Guimerà, L.A.N. Amaral, Functional cartography of complex metabolic networks. Nature **433**, 895–900 (2005). https://doi.org/10.1038/nature03288

8. E. Ravasz, A.L. Somera, D.A. Mongru, Z.N. Oltvai, A. L. Barabási, Hierarchical organization of modularity in metabolic networks. Science **297**, 1551–1555 (2002)

9. P. Erdös, A. Rényi, Publ. Math. (Debrecen) **6**, 290 (1959)

10. P. Erdös, A. Rényi, Publ. Math. Inst. Hung. Acad. Sci. **5**, 17 (1960)

11. P. Erdös A. Rényi, Bull. Inst. Int. Stat. **38**, 343 (1961)

12. D.J. Watts, S.H. Strogatz, Collective dynamics of 'small-world' networks. Nature **393**, 440 (1998). https://doi.org/10.1038/30918

13. R. Albert, A.-L. Barabási, Statistical mechanics of complex networks. Rev. Mod. Phys. **74**(1), 47–97 (2002)

14. A.-L. Barabási, R. Albert, Emergence of scaling in random networks. Science **286**, 509 (1999)

15. F. Papadopoulos, M. Kitsak, M. Ángeles Serrano, M. Boguñá, D. Krioukov, Popularity versus similarity in growing networks. Nature **489**, 537–540 (2012). https://doi.org/10.1038/nature11459

16. R.W. Benz, S.J. Swamidass, P. Baldi, Discovery of power-laws in chemical space. J. Chem. Inf. Model. **48**, 1138 (2008)

17. N. Tanaka, K. Ohno, T. Niimi, A. Moritomo, K. Mori, M. Orita, Small-world phenomena in chemical library networks: application to fragment-based drug discovery. J. Chem. Inf. Model. **49**, 2677 (2009)

18. M. Fiedler, Algebraic connectivity of graphs. Czech. Math. J. **23**(98), 298–305 (1973)

19. R. Albert, H. Jeong, A.-L. Barabási, Error and attack tolerance of complex networks. Nature **406**, 378–382 (2000)

20. M. Wawer, L. Peltason, N. Weskamp, A. Teckentrup, J. Bajorath, Structure−activity relationship anatomy by network-like similarity graphs and local structure−activity relationship indices. J. Med. Chem. **51**(19), 6075–6084 (2008)

21. E. Estrada, J.A. Rodríguez-Velázquez, Subgraph centrality in complex networks. Phys. Rev. E **71**, 056103 (2005)

22. R. Forman, Bochner's method for cell complexes and combinatorial ricci curvature. Discrete Comput. Geom. **29**, 323–374 (2003)

23. Y. Ollivier, Ricci curvature of Markov chains on metric spaces. J. Funct. Anal. **256**(3), 810–864 (2009)

24. R.P. Sreejith, K. Mohanraj, J. Jost, E. Saucan, A. Samal, Forman curvature for complex networks. J. Stat. Mech. 063206 (2016). https://doi.org/10.1088/1742-5468/2016/06/063206

25. R.P. Sreejith, J. Jost, E. Saucan, A. Samal, Systematic evaluation of a new combinatorial curvature for complex networks. Chaos Solit. Fractals **101**, 50–67 (2017)

26. W. Leal, G. Restrepo, P.F. Stadler, J. Jost, Forman-Ricci Curvature for Hypergraphs. arXiv: 1811.07825v1 [cs.DM], 19 Nov 2018

27. M.E.J. Newman, Modularity and community structure in networks. Proc. Nat. Acad. Sci. **103**(23), 8577–8582 (2006). https://doi.org/10.1073/pnas.0601602103

Chapter 4
Universality and Random Matrix Theory

Not all chemical spaces are created equal!
—Gerry Maggiora [1]

4.1 Eigenvalue Correlations

It is worth delving deeper into the eigenvalues of the matrices associated with CSNs, specifically into the correlations between these eigenvalues, because such correlations can reveal a lot about the topology of the underlying space, as well as the similarities between molecules. But as we shall see below, there are in-built correlations between the eigenvalues of even random networks. Which correlations are intrinsic to the structure of the space and the chosen similarity metric, and which of them reveal non-trivial similarity relationships between molecules? These are non-trivial questions, because as we shall see in the next chapter, not all chemical spaces are created equal [1]. But random matrix theory can provide some clues.

Random Matrix theory (RMT) was developed by Eugene P. Wigner in 1951 [2–4] to describe the correlations among the eigenvalues and eigenfunctions of complicated many-body quantum systems in nuclear physics. RMT describes statistical correlations in systems with many degrees of freedom, where the interplay between stochasticity and symmetry leads to the emergence of universal laws. Wigner considered ensembles of dynamical Hamiltonian systems with a common symmetry property, and replaced the Hamiltonian of the system by an ensemble of random Hamiltonians, to extract common generic properties of the ensemble that are consequences of the underlying fundamental symmetries. Although early applications were confined to nuclear physics, RMT has now been applied successfully in a diverse range of fields, to problems such as spectral fluctuations of atomic nuclei, as well as of complex atoms and molecules, chaotic dynamics in disordered systems, quantum phenomena in strongly interacting many-body systems, chiral symmetry breaking in quantum chromodynamics, and even quantum gravity.

The elements of the Hamiltonian matrix $\mathbf{H}$ in RMT are treated as independent random variables taken from a distribution. Freeman Dyson [5] showed that, with a

Gaussian probability distribution:

$$P_N(\mathbf{H}) \propto \exp\left\{-\frac{\beta}{2} \operatorname{Tr}.\mathbf{H}^2\right\}, \tag{4.1}$$

the symmetries of the Hamiltonian allow one to distinguish three classes of canonical ensembles, characterized by the Brody parameter β, corresponding to the different symmetries:

1. *Gaussian orthogonal ensemble* (GOE, $\beta = 1$) for systems with time-reversal invariance and rotational symmetry, where the Hamiltonian matrix is real and symmetric:

$$H_{mn} = H_{nm} = H_{mn}^* \tag{4.2}$$

2. *Gaussian unitary ensemble* (GUE, $\beta = 2$) for systems that violate time-reversal invariance, but where the Hamiltonian matrix is Hermitian:

$$H_{mn} = H_{mn}^\dagger = H_{nm}^* \tag{4.3}$$

3. *Gaussian symplectic ensemble* (GSE, $\beta = 4$) for half-integer spin systems with broken rotational symmetry that are time-reversal invariant, where the Hamiltonian matrix can be expressed in terms of Pauli spin matrices.

These Gaussian ensembles are derivable from a maximum entropy principle, using an ergodic hypothesis to relate the averages over the RMT ensembles to those for the real system. While average properties vary from system to system, the statistical fluctuations around their mean values become independent of the form of the overall spectrum in the limit where the matrix size $n \to \infty$. In this limit, the eigenvalues and corresponding eigenvectors are uncorrelated Gaussian-distributed random variables. Universal laws thus emerge from the stochasticity implied by the maximum entropy principle. RMT analysis of complex networks (and complex systems in general) can thus help separate out the system dependent part of the fluctuation properties from the random universal part, depending upon the nature of eigenvalue correlations.

enote the eigenvalues of the network by λ_i, $i = 1, \ldots, N$, in increasing order, where N is the number of nodes or network size. For $\beta = 1, 2$, or 4, we have the probability distribution:

$$P_{N\beta}(\lambda_1, \lambda_2, \ldots \lambda_N) = \prod_{j>k} |\lambda_j - \lambda_k|^\beta \prod_{i=1}^N d\lambda_i. \tag{4.4}$$

This shows that closely-spaced eigenvalues are very rare compared to widely-spaced ones – a phenomenon known as level repulsion. We can see that the repulsion of eigenvalues is stronger for larger β. The level repulsion arises from the volume element in matrix space, and it influences local but not global properties.

To obtain universal properties of the eigenvalue fluctuations, effects arising out of variations of the spectral density need to be removed by "*unfolding*" the eigenvalues; this is accomplished through a transformation $\bar{\lambda}_i = N(\lambda_i)$ to constant spectral density on average (independent of λ), where

$$\overline{N}(\lambda) = \int_{\lambda\min}^{\lambda} \rho(\lambda')d\lambda'. \tag{4.5}$$

is the average integrated density of eigenvalues. The spectrum is generally numerically unfolded by fitting it to a polynomial curve. After unfolding, the nearest-neighbor eigenvalue spacings are:

$$s_1(i) = \bar{\lambda}_{i+1} - \bar{\lambda}_i. \tag{4.6}$$

The average spacing between the unfolded nearest-neighbor eigenvalues becomes unity, independent of the system:

$$\langle s_1 \rangle = 1. \tag{4.7}$$

The *nearest-neighbor spacing distribution* (NNSD) and *next-nearest-neighbor spacing distribution* (nNNSD) are useful statistical measures to characterize eigenvalue fluctuations. The NNSD is defined as the probability distribution $P(s_1)$ of spacings between neighboring eigenvalues $s_1(i)$.

In 1957 Wigner proposed a form for the NNSD $P(s_1)$. This "Wigner surmise" takes the general form:

$$P_\beta(s_1) = a_\beta s^\beta \exp(-b_\beta s^2), \tag{4.8}$$

where the level spacing s_1 is defined in units where the local mean level spacing is unity (Eq. 4.7), a_β and b_β are constants given in terms of the gamma functions:

$$a_\beta = 2\frac{\Gamma^{\beta+1}\left(\frac{\beta+2}{2}\right)}{\Gamma^{\beta+2}\left(\frac{\beta+1}{2}\right)}$$

$$b_\beta = \frac{\Gamma^2\left(\frac{\beta+2}{2}\right)}{\Gamma^2\left(\frac{\beta+1}{2}\right)}. \tag{4.9}$$

For $\beta = 1$, 2, and 4, we have:

$$a_1 = \frac{\pi}{2}, \quad b_1 = \frac{\pi}{4} \quad \text{for GOE,}$$

$$a_2 = \frac{32}{\pi^2}, \quad b_2 = \frac{4}{\pi} \quad \text{for GUE,}$$

Fig. 4.1 Average level density $\rho(\lambda)$. The histogram shows $\rho(\lambda)$ for the Erdös-Rényi random network; the solid line shows the prediction of Wigner's semicircle law

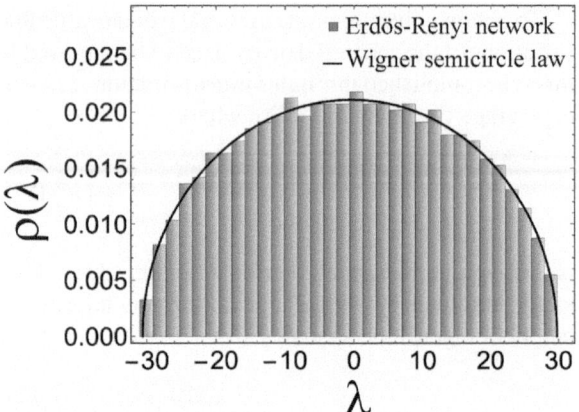

$$a_4 = \frac{262,144}{729\pi^3}, \quad b_4 = \frac{64}{9\pi} \quad \text{for GSE.} \tag{4.10}$$

The Wigner surmise is an excellent approximation to the exact form of the NNSD. It displays strong level repulsion at small spacings, dependent on β, and falling off as a Gaussian for large spacings. All three Gaussian ensembles have spectra bounded in the $n \to \infty$ limit, with length 4ζ. The average level density $\rho(\lambda)$ takes on a semicircular shape in the interval $-2\zeta \leq \lambda \leq 2\zeta$:

$$\rho(\lambda) = \left(\frac{n}{\pi\zeta}\right)\sqrt{1\left(\frac{\lambda}{2\zeta}\right)^2}. \tag{4.11}$$

This is known as Wigner's semicircle law, and Fig. 4.1 shows that it provides an excellent approximation to the average level density for the Erdös-Rényi random network.

The spectral density for scale-free networks, however, follows a triangular distribution, while those of small-world networks are more complex. For Poisson statistics, the NNSD takes the form:

$$P(s_1) = \exp(-s_1), \tag{4.12}$$

whereas for GOE:

$$P(s_1) = \frac{\pi}{2}s_1\exp\left(-\frac{\pi}{4}s_1^2\right) \tag{4.13}$$

The Brody parameter $\beta = 0$ corresponds to Poisson statistics (uncorrelated eigenvalues), and $\beta = 1$ to GOE statistics. For statistics intermediate between the two, the NNSD is given by the Brody formula:

$$P_\beta(s_1) = As_1^\beta \exp\left(-Bs_1^{\beta+1}\right), \tag{4.14}$$

where:

$$A = (1+\beta)\alpha. \tag{4.15}$$

and α is given by the semiempirical formula:

$$\alpha = \left[\Gamma\left(\frac{\beta+2}{\beta+1}\right)\right]^{\beta+1} \tag{4.16}$$

For GUE ($\beta = 2$), we have:

$$P(s_1) = \frac{32}{\pi^2}s_1^2 \exp\left(-\frac{4}{\pi}s_1^2\right), \tag{4.17}$$

while for GSE ($\beta = 4$):

$$P(s_1) = \frac{2^{18}}{3^6\pi^3}s_1^4, \exp\left(-\frac{64}{9\pi}s_1^2\right). \tag{4.18}$$

The spacings between the unfolded next-nearest-neighbor eigenvalues are given by:

$$s_2(i) = \frac{1}{2}\left(\overline{\lambda}_{i+2} - \overline{\lambda}_i\right). \tag{4.19}$$

The factor ½ makes the average next-nearest-neighbor spacing unity:

$$\langle s_2 \rangle = 1. \tag{4.20}$$

The next-nearest-neighbor spacings distribution nNNSD $P(s_2)$ for GOE and GSE is given by:

$$P(s_2) = \frac{2^{18}}{3^6\pi^3}s_2^4 \exp\left(-\frac{64}{9\pi}s_2^2\right). \tag{4.21}$$

Jalan and Bandyopadhyay [6, 7] numerically investigated the NNSD and nNNSD of eigenvalues of the adjacency matrix for several random, scale-free, and small-world networks, and found that all these distributions follow GOE statistics, with very similar NNSD and nNNSD profiles. This universality was attributed to a certain minimum amount of intrinsic randomness introducing correlations among neighboring eigenvalues, in all these networks. These authors also studied the yeast protein–protein interaction network and found that it too follows GOE statistics.

Another useful RMT statistic is the ratio of consecutive spacings between neighboring eigenvalues. The advantage of this metric is that it does not depend upon the local density of states. It can be shown that the ratio of consecutive spacings between raw eigenvalues λ_i is identical to the corresponding ratio between unfolded eigenvalues:

$$r_i = \frac{\lambda_{i+2} - \lambda_{i+1}}{\lambda_{i+1} - \lambda_i} = \frac{\overline{\lambda}_{i+2} - \overline{\lambda}_{i+1}}{\overline{\lambda}_{i+1} - \overline{\lambda}_i}. \tag{4.22}$$

This means that determining its distribution does not require an unfolding procedure, which is often difficult to perform in practice. This makes the *ratio distribution* a more trustworthy metric for short-range correlations. The ratio distribution for GOE is given as:

$$P(r) = \frac{27}{8} \frac{r + r^2}{\left(1 + r + r^2\right)^{\frac{5}{2}}}. \tag{4.23}$$

Figure 4.2 shows that the NNSD and the eigenvalue spacing ratio distributions are broadly similar for the adjacency and Laplacian matrices of various CSNs [8, 9], and follow GOE rather than Poisson statistics. Mishra, Raghav and Jalan [10] found that the eigenvalue spacing ratios distributions of the adjacency matrices for various small-world, Erdös-Rényi and other random networks follow GOE statistics. But adding increasing amounts of disorder to the diagonal terms of the adjacency matrices resulted in a gradual transition to Poisson statistics.

Measures like NNSD and nNNSD encode the short-range correlations among the matrix eigenvalues. Metrics such as the level *number variance* Σ^2 and the *spectral rigidity* Δ^3 reflect long-range correlations. Denoting the number of distinct eigenvalues in the interval $[\overline{\lambda}_s, \overline{\lambda}_s + L]$ on the unfolded scale by $\rho(L, \overline{\lambda}_s)$, the level number variance is given by:

$$\Sigma^2(L) = \left\langle \left(\rho(L, \overline{\lambda}_s)\right)^2 \right\rangle - \left\langle \rho(L, \overline{\lambda}_s)^2 \right\rangle. \tag{4.24}$$

Note that as a consequence of unfolding, the local density of states is constant:

$$\left\langle \rho(L, \overline{\lambda}_s) \right\rangle = L. \tag{4.25}$$

The spectral staircase function $\eta(x)$ represents the average integrated density of eigenvalues:

$$\eta(x) = \sum_{-\infty}^{x-L} \rho(L, x'). \tag{4.26}$$

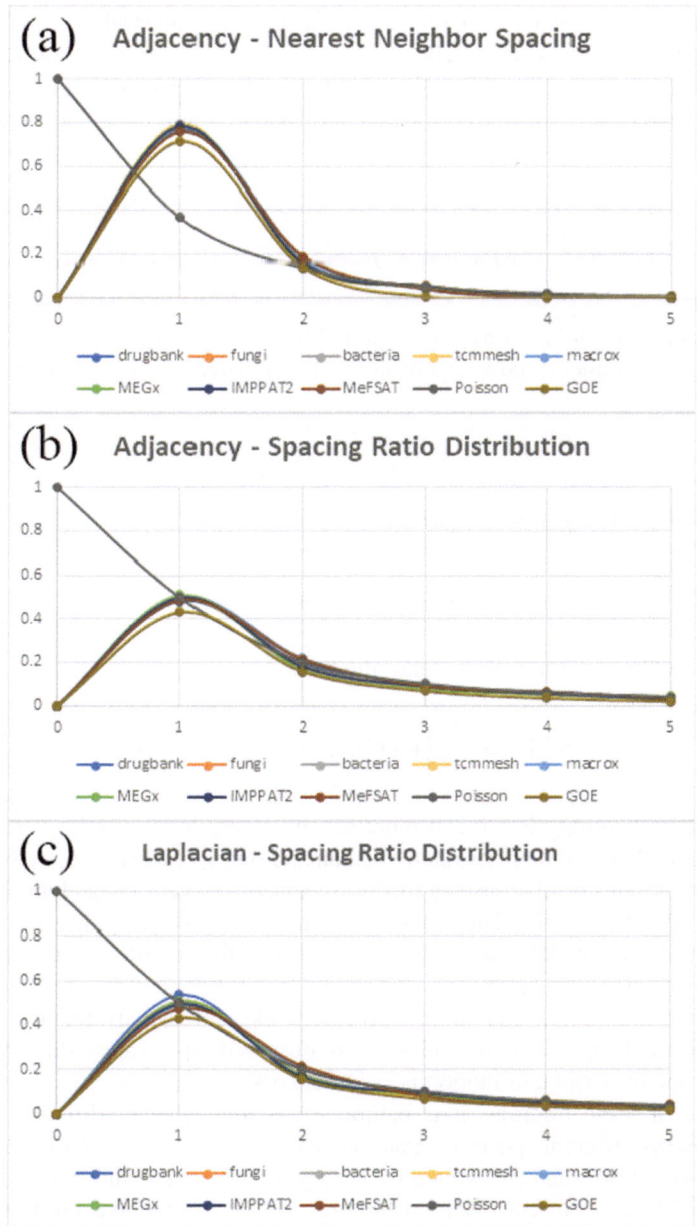

Fig. 4.2 **a** Nearest neighbor spacing distribution of the adjacency matrix, and eigenvalue spacing ratio distribution for the **b** adjacency and Laplacian matrices of various chemical libraries from Ref. [9]—Drugbank (2466 molecules), IMPPAT2 (17,915 molecules), Macrox (4306 molecules), MeFSAT (1829 molecules), MEGx (6458 molecules), NP atlas bacteria (12,505 molecules), NP atlas fungi (19,966 molecules), TCM mesh (10,127 molecules), as well as for a Poisson distribution and a Gaussian orthogonal ensemble. Generated by the author from data in Refs.8 and 9

It is the cumulative number of energy levels in the interval $[-\infty, x]$ on the unfolded, dimensionless scale. The spectral rigidity Δ^3, introduced by Dyson and Mehta [11], measures the "stiffness" of the eigenvalue spectrum, or the deviation of the best straight line fit from the spectral staircase function $\eta(x)$ in a finite spectral interval L:

$$\Delta^3(L) = \frac{1}{L}\left\langle \min_{a,b} \int_x^{x+L} [\eta(x) - ax - b]^2 d\lambda' \right\rangle_x, \qquad (4.27)$$

where a and b are fitting constants obtained from a least-square fit. For a Poisson distribution, with uncorrelated eigenvalues, $\Delta^3(L)$ depends linearly on L:

$$\Delta^3(L) = \frac{L}{15}. \qquad (4.28)$$

while for GOE, $\Delta^3(L)$ shows logarithmic dependence on L:

$$\Delta^3(L) \sim \frac{1}{\pi^2} \log L. \qquad (4.29)$$

For GOE, the spectral rigidity takes the form:

$$\Delta^3(L) = \frac{1}{\pi^2}\left\{ \log(2\pi L) + \gamma - \frac{5}{4} - \frac{\pi^2}{8} \right\}. \qquad (4.30)$$

The spectral rigidity is a much more sensitive and discriminating statistic than NNSD and nNNSD, as demonstrated by Jalan and Bandyopadhyay [6, 7], who investigated long-range correlations among the eigenvalues of various random, scale-free, and small-world networks. They found that while the spectral rigidity Δ^3 of random networks followed the RMT prediction of Eq. (4.29) (linear with respect to log L with slope $1/\pi^2$) even at very large value of L, scale-free networks followed this behavior to a smaller value of L, and small-world networks did so only for low values of L, beyond which significant deviations were observed. Spectral rigidity can thus be used to probe the degree of randomness in networks.

In light of these findings, it is natural to wonder whether different chemical spaces possess different spectral signatures that can be distinguished by RMT analysis. This question was investigated by Kothiyal, et al. [12], who studied the RMT statistics of three different CSNs: a focused library of potential inhibitors of the enzyme Glutaminase C, and two diversified CSNs: one consisting of traditional Chinese medicine (TCM) and a set of naturally occurring plant-based anti-cancer agents (NPACT) from the ZINC database. By varying the similarity threshold, and thereby the edge density of the networks, it was found that while NNSD, nNNSD and spacing ratio distribution did not distinguish between different chemical spaces, all of which followed GOE predictions throughout the range of edge densities, the spectral rigidity followed GOE statistics only at moderate edge density, and was sensitive to variations in the global network structure as well as to the chemical space (Fig. 4.3).

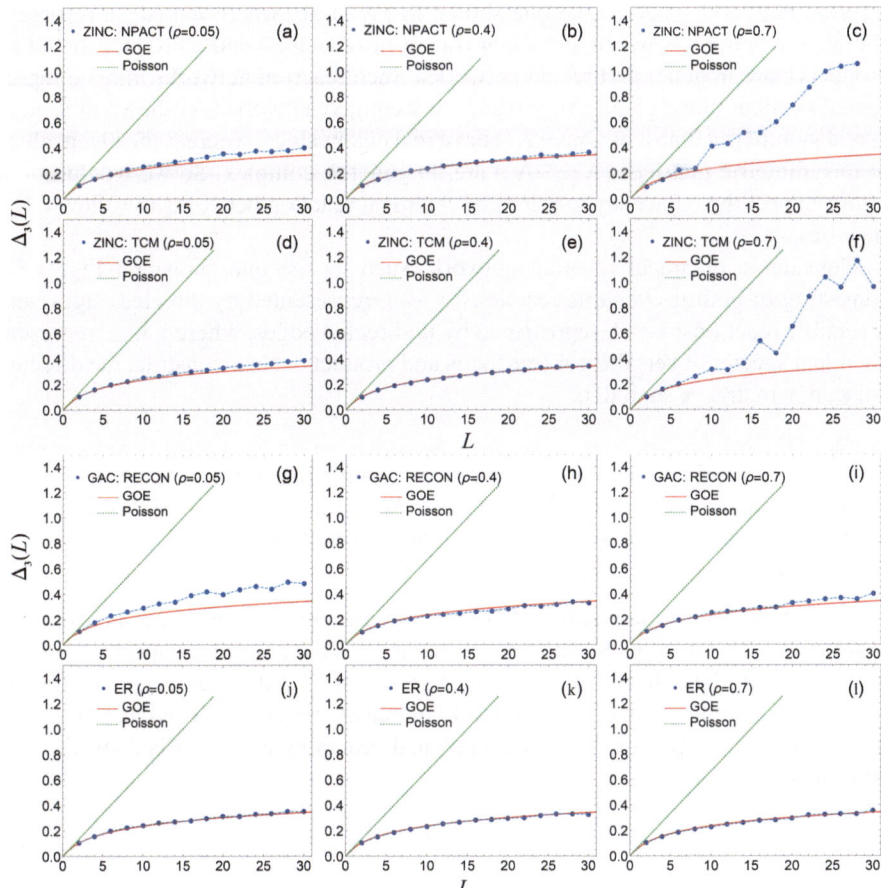

Fig. 4.3 Variation of spectral rigidity $\Delta^3(L)$ with edge density **a–c** for the Naturally Occurring Plant-based Anti-cancer Compound-Activity-Target (NPACT) library of the ZINC database, **d–f** for the traditional Chinese medicines (TCM) library of the ZINC database, **g–i** for a focused library of GAC inhibitors, and **j–l** for the Erdös-Rényi random network. Dots connected with dashed (blue) curve correspond to the empirical results. The solid (red) curve and dotted (green) curve in each case are the GOE and Poisson results. Adapted with permission from Ref. [12]. © 2022, Springer Nature. All rights reserved

4.2   RMT for Chemical Reaction Networks

Unlike similarity networks that are undirected networks with $A_{ij} = A_{ji}$, chemical reaction networks, such as the one shown in Fig. 1.3c, where the edges represent chemical reactions between a reactant (or set of reactants) and a product (or set of products), are in general, directed networks. Such reaction networks may be represented as either unweighted ($A_{ij} = 0$ or 1) or weighted networks. As shown in Sect. 3.2, real symmetric matrices ($A_{ij} = A_{ji}$) have real eigenvalues, whereas the eigenvalues of unsymmetric matrices ($A_{ij} \neq A_{ji}$) are, in general, complex. So we would not, a priori, expect the adjacency matrices of chemical reaction networks to follow GOE statistics.

Consider a chemical reaction network, such as the one shown in Fig. 1.3c, consisting of both irreversible reactions: $j \rightarrow l$, represented by directed edges, and reversible reactions: $j \leftrightarrow l$, represented by undirected edges, where j, l, ... represent chemical species or sets thereof (reactants and products). Now construct the directed adjacency matrix **A** such that:

- $A_{ii} = 0$,
- $A_{jl} = 1$ if j and l are related through the reversible reaction $j \leftrightarrow l$,
- $A_{jl} = i$ if j and l are related through the irreversible reaction $j \rightarrow l$, and
- $A_{jl} = -i$ if j and l are related through the irreversible reaction $j \leftarrow l$ (where $i = \sqrt{-1}$).

Thus **A** is a Hermitian matrix [13], and has real eigenvalues. It can be expected to follow GUE statistics. We can expect a change in the eigenvalue spectrum of this directed Hermitian adjacency matrix as the fraction of undirected edges (reversible reactions in the chemical reaction network) increases from 0 to 1. This can be effected by raising the temperature T, since chemical reactions are described by the rate constants:

$$k_{jl} = F_{jl} \exp\left\{-\frac{\Delta E_{jl}}{kT}\right\},   \qquad (4.31)$$

where ΔE_{jl} is the activation energy for the reaction $j \rightarrow l$, F_{jl} is the Arrhenius factor, and k is the Boltzmann constant. We can also consider the weighted network of rate constants **H**, such that:

$$H_{jl} = \frac{1}{2}(k_{jl} + k_{lj}) + \frac{i}{2}(k_{jl} - k_{lj})   \quad \text{for } j \neq l   \qquad (4.32)$$

and

$$H_{jj} = 0.$$

We see that Eq. (4.3) is satisfied, so that **H** is a Hermitian matrix. For a fully reversible reaction, $k_{jl} = k_{lj}$ and $H_{jl} = k_{jl}$ is real, for all j, l. Thus, in the limit T $\rightarrow$

∞, we have $|k_{jl} - k_{lj}| \to 0$, for all j, l, and the matrix **H** becomes real symmetric, with real eigenvalues. As the temperature is raised, we would thus expect to see a gradual transition from GUE to GOE statistics [14].

In the general case, where there may be multiple reactants and multiple products, if each chemical species is represented by a node, we need to go beyond graphs or networks and deal with simplices or hypergraphs. The connections in such objects can no longer be represented by adjacency matrices, but by tensors which can have any number of dimensions.

4.3 RMT for Feature Networks

RMT analysis of feature networks provides an elegant way to tackle the issue of feature selection, considered as a denoising problem. Lee, Brenner and Colwell [15] used RMT for feature selection in protein–ligand binding affinity prediction. RMT analysis of correlations among molecular descriptors was employed to eliminate statistically insignificant correlations, and the remaining, selected descriptors were then used in a classification algorithm to predict whether or not a ligand will bind to a given protein receptor.

We have already seen in Sect. 1.6 that most available descriptors are irrelevant for a particular problem, such as binding to a specific target, and only add noise to a QSAR model. QSAR generally looks at correlations between the descriptors and the activity or binding affinity, but the correlations of the descriptors with one another also contain valuable information about binding when we compare randomly chosen ligands to those specifically selected for their binding affinity to a given target. We have seen in Sect. 4.1 that even random networks exhibit correlations among their eigenvalues. For a set of randomly chosen ligands, many descriptors may show some degree of inter-correlation, but for ligands that bind to the same protein target, those descriptors that are strongly correlated with the binding affinity will also show strong inter-correlation among themselves by virtue of their common correlation with the binding affinity (homophily principle for correlation networks). The correlation threshold distinguishing between chance inter-correlation and that due to ligand binding is given by RMT.

For samples drawn from a Gaussian distribution that has zero mean and unit variance, the distribution of eigenvalues λ of the correlation matrix is given by the Marčenko–Pastur (MP) distribution [16] of RMT:

$$\rho(\lambda) = \frac{\sqrt{\left[\left(1 + \sqrt{\lambda}\right)^2 - \lambda\right]_+ \left[\lambda - \left(1 - \sqrt{\lambda}\right)^2\right]_+}}{2\pi\gamma\lambda}, \tag{4.33}$$

where γ is a parameter describing the degree of sampling of the dataset. The probability of eigenvalues larger than a threshold value of $\left(1 + \sqrt{\lambda}\right)^2$ for a random matrix in the absence of a signal is close to zero. Thus, eigenvalues above this threshold correspond to statistically significant correlations. For a random set of n molecules, represented by a mean-centered, standard deviation-scaled data matrix **A** of Boolean fingerprints, the eigenvalue distribution of the n × n covariance matrix $\mathbf{A}^T\mathbf{A}/n$ (Eq. 2.58a) of molecular fingerprints can be expected to follow the MP distribution, which therefore provides a null hypothesis for chance correlations. For fingerprints of pharmacologically similar molecules binding to the same protein target, part of the eigenvalue spectrum will still follow the MP distribution. But only eigenvectors of the covariance matrix with eigenvalues larger than a threshold value, derivable from the MP distribution, correspond to statistically significant chemical features relevant for ligand binding to the given receptor. This MP threshold thus serves to distinguish chance correlations from those caused by binding to that receptor. Computing the common substructure among the ligands that are most similar to these eigenvectors identifies the structural motif responsible for ligand binding. The RMT approach was found to maximize the information extracted from inter-feature correlations, while avoiding overfitting, and thereby to outperform other methods in predicting protein–ligand binding affinity.

References

1. G. Maggiora, On outliers and activity cliffs: why QSAR often disappoints. J. Chem. Inf. Model. **46**(4), 1535 (2006). https://doi.org/10.1021/ci060117s
2. E.P. Wigner, Characteristic vectors of bordered matrices with infinite dimensions. I. Ann. Math. **62**(3), 548–564 (1955)
3. T. Guhr, A. Müller-Groeling, H.A. Weidenmüller, Random-matrix theories in quantum physics: common concepts. Phys. Rep. **299**, 189–425 (1998)
4. M.L. Mehta, *Random Matrices* (Elsevier, Amsterdam, 2004)
5. F.J. Dyson, J. Math. Phys. **3**, 140 (1962)
6. S. Jalan, J.N. Bandyopadhyay, Random matrix analysis of complex networks. Phys. Rev. E. E **76**, 046107 (2007). https://doi.org/10.1103/PhysRevE.76.046107
7. J.N. Bandyopadhyay, S. Jalan, Universality in complex networks: random matrix analysis. Phys. Rev. E. E **76**, 026109 (2007). https://doi.org/10.1103/PhysRevE.76.026109
8. R.P. Vivek-Ananth, A.K. Sahoo, S.P. Baskaran, A. Samal, Scaffold and structural diversity of the secondary metabolite space of medicinal fungi. https://doi.org/10.1101/2022.09.25.509364
9. R.P. Vivek-Ananth, K. Mohanraj, A.K. Sahoo, A. Samal, IMPPAT 2.0: an enhanced and expanded phytochemical atlas of Indian medicinal plants. ACS Omega **8**(9), 8827–8845 (2023). https://doi.org/10.1021/acsomega.3c00156
10. A. Mishra, T. Raghav, S. Jalan, Eigenvalue ratio statistics of complex networks: disorder versus randomness. Phys. Rev. E. E **105**, 064307 (2022). https://doi.org/10.1103/PhysRevE.105.06430
11. F.J. Dyson, M.L. Mehta, J. Math. Phys. **4**, 701 (1963)
12. M. Kothiyal, S. Kumar, N. Sukumar, Investigation of chemical space networks using graph measures and random matrix theory. J. Math. Chem. **60**, 891–914 (2022). https://doi.org/10.1007/s10910-022-01341-y

13. T. Peron, B. Messias F. de Resende, F.A. Rodrigues, L.F. Costa, J.A. Méndez-Bermúdez, Spacing ratio characterization of the spectra of directed random networks. Phys. Rev. E **102**, 062305 (2020). https://doi.org/10.1103/PhysRevE.102.062305
14. S. Sowmya, N. Sukumar, S. Kumar, unpublished
15. A.A. Lee, M.P. Brenner, L.J. Colwell, Predicting protein–ligand affinity with a random matrix framework. Proc. Nat. Acad. Sci. **113**(48), 13564–13569 (2016)
16. V.A. Marčenko, L.A. Pastur, Distribution of eigenvalues for some sets of random matrices. Math. USSR-Sbornik **1**, 457 (1967)

Chapter 5
Mapping and Navigating Chemical Space Networks

Prediction is difficult, especially when dealing with the future. (variously attributed to Neils Bohr, Samuel Goldwyn and Yogi Berra).

5.1 *k*-NN and *k*-Means

ML techniques follow one of three broad paradigms: supervised, unsupervised and reinforcement learning (with some intermediate cases such as semi-supervised and self-supervised learning). *Supervised learning* is where there is a training set of molecules with labeled data, *i.e.* whose target activities have been experimentally measured. The ML model is first trained on this set, and the trained model is then used to predict the activities of other molecules that are not present in the training set. In *unsupervised learning*, there is only unlabeled data—we do not have a training set with known activities—and the data are classified based solely on the distribution and statistical properties of the dataset. The k-nearest neighbor (k-NN) algorithm is perhaps the simplest supervised machine learning technique that implicitly constructs a map of the immediate neighborhood around every molecule in chemical space, by building a directed regular graph of constant in-degree k. The construction of this k-regular graph from a similarity matrix like that of Table 1.3a is equivalent to choosing a local similarity threshold around each molecule such that its in-degree is exactly k, as shown in Table 5.1.

k-NN is one of the most widely used ML methods for both classification and regression, and is based on the principle that similar molecules, i.e. those belonging to the same class in a classification problem, should lie close together in chemical space. In k-NN classification, when the available data includes only binary or categorical labels, and we wish to predict the labels for the test set, new molecules are classified according to the closest k training molecules in chemical space. The first step in the procedure is to choose the value of k; choosing an appropriate value is essential to avoid overfitting. The algorithm then computes the Euclidian distances in chemical space between the new molecule and all the training molecules, and then finds the

N. Sukumar, *Navigating Molecular Networks*, SpringerBriefs in Materials, https://doi.org/10.1007/978-3-031-76290-1_5

Table. 5.1 Incidence matrix for 3-nearest neighbor (3-NN) network constructed from the Tanimoto similarity matrix of Table. 1.3a

Node	3 Nearest Neighbors		
1	2	4	12
2	1	9	12
3	5	12	13
4	6	7	8
5	3	12	13
6	4	7	11
7	4	6	8
8	4	6	7
9	1	2	10
10	1	2	9
11	4	6	7
12	1	2	3
13	3	5	12

class labels or activities of its k nearest neighbors. In k-NN classification, the class label of the new molecule is assigned by majority vote of its k nearest neighbors (with k odd). In k-NN regression, where the training set has numerical properties and we wish to predict the numerical values of the target properties for the test set, the activity of the new test molecule is determined by averaging the activities of its k nearest neighbors in chemical space.

k-means is an unsupervised ML algorithm for clustering, where the dataset is partitioned into k clusters based on their features. It is equivalent to decomposition of a weighted network into communities. Each cluster or community is represented by its centroid in chemical space, and the algorithm tries to minimize the inter-cluster variance. The first step is again to choose an appropriate value for the parameter k. The algorithm then starts with an initially random assignment of the k centroids, and computes the Euclidian distance of each molecule to all the cluster centroids. Each molecule's class membership is then assigned to its nearest centroid, and the centroid of each cluster is updated by taking the mean of the data points associated with it. These two steps are repeated, minimizing the sum of the squares of the Euclidian distances of each molecule to its cluster centroid, until convergence is achieved, and no further optimization is possible. The k-means algorithm is fast and simple, but due to the initial random assignment of centroids, it does not yield the same result with each run, and convergence is not guaranteed. Figure 5.1a shows the result of k-means clustering, with $k=3$, for the dataset of Fig. 2.1.

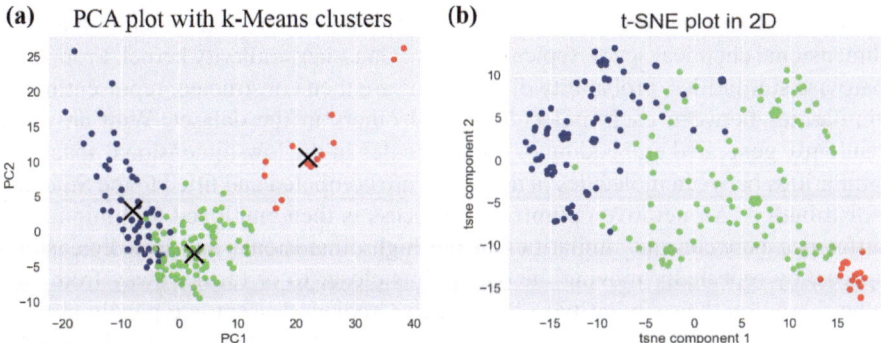

Fig. 5.1 (**a**) k-means clustering, with $k = 3$, and (**b**) t-SNE map, for the dataset of Fig. 2.1

5.2 Visualizing Chemical Space Networks

The *Kohonen map* or Self-Organizing Map (SOM) [1] is a clustering algorithm, based on unsupervised Artificial Neural Networks (ANN), and employing a process of competitive learning. The network consists of several layers of nodes or "neurons." Each layer projects the input, which it receives from the layer immediately below it, to a layer directly above it. The competition is induced by inhibitory connections between the neurons in a layer. Neurons within a layer are organized into inhibitory clusters, in which the neurons compete with each other to be activated. Each neuron is connected via excitatory connections to the layers immediately above and below it, and with inhibitory connections to neurons within the competition layer. A weight vector is associated with each connection, and the SOM discovers the underlying structure in the data by repeatedly moving its neurons closer to the data points; the neuron whose weight vector is closest to that of the input vector is the winner in the race to be activated. As a result, the neurons learn to organize themselves, such that the SOM evolves to take on the shape of the input data. The SOM is thus a topology-preserving map, providing a way to represent multi-dimensional data in a low dimensional space, which makes it useful for data visualization, and to show the relations between multiple variables. The limitations are that SOMs are computationally expensive, and can generate potentially inconsistent solutions.

In Sect. 2.2 we discussed PCA as a dimensionality reduction and data visualization technique, but because of its linear nature, PCA is not ideal for visualizing highly nonlinear data. Two of the most popular data visualization methods in ML are t-SNE and UMAP, both of which are nonlinear, graph-based algorithms for dimensionality reduction. Both use graph layout algorithms to map a high dimensional graph into a lower dimensional space while retaining its structure.

t-SNE (t-distributed Stochastic Neighbor Embedding) [2] is an unsupervised algorithm that works by generating a probability distribution from the similarities between molecules in the high-dimensional space, and mapping this probability distribution onto a lower-dimensional space, while preserving the relationships between the molecules. Pairwise similarities between molecules are first computed in the high-dimensional chemical space, typically using a Gaussian similarity kernel. From these pairwise similarities, probability distributions are then constructed, representing the similarities between each molecule and all others in the dataset. With an initial randomly generated embedding of the molecules in the low-dimensional space, the similarities between molecules in this space are computed and fitted to the Student's t-distribution. An iterative optimization process is then employed to minimize the difference between the similarities in the high-dimensional and low-dimensional spaces, by repeatedly moving the data points closer to or further away from each other based on their similarity to each other. Gradient descent is typically used for optimization, which continues until convergence is achieved. The molecules are then visualized in the low-dimensional space. The optimized embedding determines the positions of the molecules in this space. t-SNE captures the local nonlinear structure in the high dimensional data: if two points are close together in the original vector space, they will also be close together in the low-dimensional embedding space. Figure 5.1b shows a t-SNE map for the dataset of Fig. 2.1.

Since t-SNE is a stochastic method, multiple runs on the same dataset can give maps that look different. Since the optimization is performed pair-wise, t-SNE is very slow compared to other dimensionality reduction techniques, and does not scale well with large datasets. Key tunable parameters in t-SNE are the perplexity (an estimate of the number of close neighbors or average degree, which controls the balance between local and global clustering), the number of iterations and the learning rate.

UMAP (Uniform Manifold Approximation and Projection) [3] works by first constructing a high dimensional graph representation of the data, and the layout of a low-dimensional graph is then optimized to be as structurally similar as possible, while preserving the global and local topology of the data. The first step is again to compute the pairwise similarities between molecules, based on which a fuzzy simplicial set is constructed representing the local similarity neighborhood around each molecule, taking into consideration the local density in chemical space. UMAP then compresses the graph through an optimization process to generate a low-dimensional representation preserving the neighborhood relationships. Starting with an initially random low-dimensional embedding, the difference between the pairwise similarities in the high- and low-dimensional spaces is minimized using stochastic gradient descent. A graph representation is then constructed based on the low-dimensional embeddings, so as to preserve both the neighborhood relationships between molecules and the global structure of the data. A refinement step attempts to improve the embedding by adjusting the positions of the molecules in the low-dimensional space based on the local connectivity and graph structure. The optimization and refinement steps are repeated iteratively until convergence.

The two key tunable parameters in UMAP are the number of nearest neighbors used to construct the initial high-dimensional graph and the minimum distance between points in the low-dimensional space. The number of nearest neighbors determines the balance between local and global structure in the final embedding. A low value will emphasize the local structure, while a high value will emphasize the global structure at the cost of losing fine details. The minimum distance between points controls the extent of clumping of nearby points. A low value will generate a tightly packed embedding, while a large value will generate a looser embedding that focuses on preservation of the broad topological structure. UMAP offers significant time and computation cost savings compared to t-SNE, especially when dealing with large datasets. It also strikes a balance between preserving local versus global structure in the data, and thus does a better job at preserving global structure compared to t-SNE.

5.3 Model Applicability Domain and Scaffold Hopping

Model *domain of applicability* is the training space on which the model has been constructed, and for which it can be applied to make predictions on new molecules [4]. The predictions of a QSAR/QSPR model are reliable only for molecules within its applicability domain. While interpolation within the applicability domain of a model is the basis of QSAR/QSPR and much of ML regression, extrapolation outside the applicability domain is fraught with peril. While some amount of extrapolation might be considered acceptable, predictions for molecules well outside the applicability domain cannot be trusted. Methods of estimating a model's applicability domain [5] include range-based (whether or not the test molecule falls within the range of descriptors or principal components of the training set), distance-based (estimated by computing the distance of the test molecule from the training set), geometrical (based on determining the smallest convex hull covering the training set—with the disadvantages that this cannot identify empty spaces within the convex hull, and that the computation time grows with the number of molecules in the training set), and probability density distribution based (considered the most robust, but also more computationally intensive).

Scaffold hopping is the ability of a model to generalize across diverse structural scaffolds. A local model is one with an applicability domain restricted to one class of structures. A generalizable model, on the other hand, because it has been trained on a broader training set or because of the nature of molecular representation employed, is capable of 'hopping' from one kind of structural scaffold to a very different one. 2D structural fingerprints, Boolean arrays where bits represent the presence or absence of specific structural features, are generally not capable of scaffold hopping because they encode information in substructural keys. 3D descriptors constructed from the distance matrix can potentially overcome this limitation of 2D fingerprints, to produce more global QSAR models. Local molecular surface property descriptors [6, 7] that do not directly encode the chemical constitution of a molecule, can also lead to models capable of scaffold hopping, especially for bioactivities primarily governed by non-covalent interactions. Such descriptors include matrix eigenvalues, van der Waals surface histograms, local maxima and local minima of properties such as the electrostatic potential or the average local ionization energy (a quantum chemical descriptor related to the Lewis acidity or basicity or how tightly a given region on the molecular surface holds on to its electrons).

5.4 Violation of the Similarity Principle—Activity Cliffs

QSAR assumes a smooth relationship between biological activity and the molecular descriptors employed in a model. Yvonne Martin and colleagues [8] challenged the validity of this principle, and found that molecules highly similar to active ones across several high-throughput screening (HTS) assays, only had a 30% probability of being active themselves. This violation of the similarity principle (Sect. 2.1) can be attributed to inadequacies in the molecular representation and/or the similarity assessment metric used in the model, as well as to the diverse ways drugs interact with molecules in the body (different modes of action). Maggiora [9] termed such discontinuities in the structure–activity landscape as "Activity cliffs," namely abrupt changes in biological activity caused by minor structural variations, emphasizing that "Not all chemical spaces are created equal!".

Guha and van Drie [10] introduced the Structure–Activity Landscape Index (SALI), a quantitative measure of activity cliffs in a structure–activity relationship:

$$\text{SALI}_{i,j} = \frac{y_i - y_j}{1 - S_{ij}}, \tag{5.1}$$

where y_i and y_j are the biological activities of molecules i and j, respectively, and S_{ij} is the structural similarity coefficient between them. Steep activity cliffs associated with high SALI values are problematic for QSAR, but at the same time they are pivotal for lead optimization, offering the potential for potency enhancement of a drug candidate through minor structural tweaks. SALI defines a directed graph (Fig. 5.2), with each edge pointing from the less potent to the more potent molecule of a pair of structurally similar molecules.

The SALI curve [11] plots the fraction S(X) of true minus false predictions of SALI edges versus the normalized activity difference threshold (X) for edge detection; it counts how many orderings of SALI edges the model correctly predicts. The ability of a model to correctly identify the steepest activity cliffs is given by S(1), while S(0) quantifies its ability to capture all SALI edges. The integral of the SALI curve (SCI) [12] should be close to 1 for a useful QSAR model.

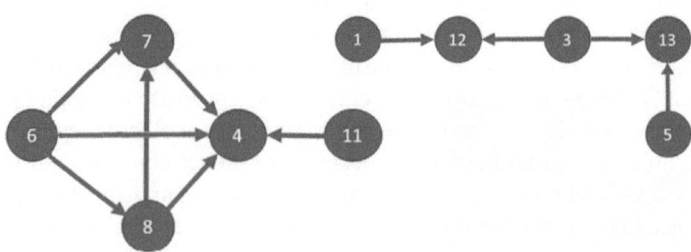

Fig. 5.2 Structure–activity landscape index (SALI) graph for the molecules in Table. 1.3a at the Tanimoto similarity threshold of 12, such that any pair of nodes with SALI ≥ 12 is connected by a directed edge. The arrows point from the less potent to the more potent molecule of each pair

While activity cliffs are valuable for lead optimization, many of them could well be outliers or noise in the data [13]. Often activity cliffs are not found in isolation but are parts of larger structures [14]. Efforts to predict [15, 16] or manage activity cliffs focus on identifying and removing *activity cliff generators* (molecules likely to show activity cliffs with other molecules in a biological assay) from QSAR training sets [13, 16]. However, this approach risks losing predictive power and limiting model applicability [17]. Achieving a balance between predictive ability and robustness of a model and preserving its applicability domain, and between compound selectivity and promiscuity [14], are essential for practical QSAR application.

References

1. T. Kohonen, Self-organized formation of topologically correct feature maps. Biol. Cybern. **43**(1), 59–69 (1982). https://doi.org/10.1007/bf00337288
2. L. Van der Maaten, G. Hinton, Visualizing data using t-SNE. J. Mach. Learn. Res. **9**, 11 (2008)
3. L. McInnes, J. Healy, J. Melville, Umap: uniform manifold approximation and projection for dimension reduction. arXiv:1802.03426 (2018)
4. J.S. Jaworska, M. Comber, C. Auer, C.J. Van Leeuwen, Summary of a workshop on regulatory acceptance of (Q)SARs for human health and environmental endpoints in Setubal, 2002. Environ. Health Persp. **111**(10), 1358–1360 (2003). https://www.jstor.org/stable/3435408
5. N. Nikolova, J. Jaworska, Approaches to measure chemical similarity—a review. QSAR Comb. Sci. **22**, 1006 (2003). https://doi.org/10.1002/qsar.200330831
6. C.E. Whitehead, C.M. Breneman, N. Sukumar, M.D. Ryan, Transferable atom equivalent multi-centered multipole expansion method. J. Comp. Chem. **24**, 512–529 (2003). https://doi.org/10.1002/jcc.10240
7. N. Sukumar, S. Das, M. Krein, C.M. Breneman, Q. Luo, R. Godawat, S. Garde, I. Vitol, K.P. Bennett, Molecular descriptors for biological systems, in *Computational Approaches in Cheminformatics and Bioinformatics*, eds. by R. Guha, A. Bender (John Wiley, Hoboken, NJ, 2012), pp. 107–144.
8. Y.C. Martin, J.L. Kofron, L.M. Traphagen, Do structurally similar molecules have similar biological activity? J. Med. Chem. **45**, 4350–4358 (2002). https://doi.org/10.1021/jm020155c
9. G. Maggiora, On outliers and activity cliffs: why QSAR often disappoints. J. Chem. Inf. Model. **46**(4), 1535 (2006). https://doi.org/10.1021/ci060117s
10. R. Guha, J. Van Drie, Structure–activity landscape index: identifying and quantifying activity cliffs. J. Chem. Inf. Model. **48**, 646–658 (2008). https://doi.org/10.1021/ci7004093
11. R. Guha, J. Van Drie, Assessing how well a modeling protocol captures a structure—activity landscape. J. Chem. Inf. Model. **48**(8), 1716–1728 (2008). https://doi.org/10.1021/ci8001414
12. N.C. LeDonne Jr., K. Rissolo, J. Bulgarelli, L. Tini, Use of structure-activity landscape index curves and curve integrals to evaluate the performance of multiple machine learning prediction models. J. Cheminformatics **3**, 7 (2011). https://doi.org/10.1186/1758-2946-3-7
13. J. Pérez-Villanueva, O. Méndez-Lucio, O. Soria-Arteche, J.L. Medina-Franco, Activity cliffs and activity cliff generators based on chemotype-related activity landscapes. Mol. Divers. **19**, 1021–1035 (2015). https://doi.org/10.1007/s11030-015-9609-z
14. D. Stumpfe, J. Bajorath, Exploring activity cliffs in medicinal chemistry. J. Med. Chem. **55**, 2932–2942 (2012). https://doi.org/10.1021/jm201706b
15. K. Heikamp, X. Hu, A. Yan, J. Bajorath, Prediction of activity cliffs using support vector machines. J. Chem. Inf. Model. **52**, 2354–2365 (2012). https://doi.org/10.1021/ci300306a
16. V. Namasivayam, J. Bajorath, Searching for coordinated activity cliffs using particle swarm optimization. J. Chem. Inf. Model. **52**, 927–934 (2012). https://doi.org/10.1021/ci400597d

17. M. Cruz-Monteagudo, J.L. Medina-Franco, Y. Pérez-Castillo, O. Nicolotti, M. Natália, D.S. Cordeiro, F. Borges, Activity cliffs in drug discovery: Dr Jekyll or Mr Hyde? Drug Disc. Today **19**(8), 1069–1080 (2014). https://doi.org/10.1016/j.drudis.2014.02.003

Chapter 6
Growing the Network—Generative AI

Networking is more about farming than it is about hunting. It's about cultivating relationships.
—Ivan Misner [1]

6.1 Genetic Algorithms

We have discussed some feature selection techniques in earlier chapters. One problem with selecting or dropping features one at a time is that it does not select sets of features that have a synergistic effect—features that may not individually be highly correlated with the target activity, but that predict well in combination. First introduced by Holland [2], *Genetic Algorithms* (GA) overcome this limitation and are now one of the most popular feature selection techniques in QSAR. GAs are a generative algorithm that, in addition to their use in feature selection [3], are also widely employed to generate new molecular structures and novel molecules; so they form a smooth transition and a gentle introduction to the main contents of this chapter. Generative algorithms attempt to tackle the long-standing dream of QSAR, namely inverse QSAR, or inverting the mapping between molecular structure and activity, enabling one to design molecules with a specific biological activity and/or within a specific range of properties.

GAs represent a class of stochastic optimization methods, that mimic how a population of chromosomes evolves through natural selection. When employed for feature selection, the GA searches the space of molecular descriptors in a random, but directed manner, by mixing and matching groups of descriptors (Table 6.1). Each individual in a population represents a group of descriptors, in the form of a binary string, called a chromosome. The length of the chromosome (i.e. the number of genes in it) is the number of descriptors in the model pool. If a particular descriptor is selected in a model, the corresponding gene has the value 1, and if it is not selected, it has the value 0. When employed for generating new molecular structures, the search is performed in chemical space by mixing and matching substructures in a molecule. Each chromosome in the population now represents a set of substructures.

N. Sukumar, *Navigating Molecular Networks*,
SpringerBriefs in Materials, https://doi.org/10.1007/978-3-031-76290-1_6

If a substructure is present in the molecule, the corresponding gene has the value 1; if not, it is 0.

A set of chromosomes makes up the population. The GA is initialized with an initial population of chromosomes. The processes of cross-over and random muta-tion (Fig. 6.1), applied to the parent population, give rise to a new generation of chromosomes. In each generation, the fitness of the population is estimated using a fitness function that allows only the fittest chromosomes to "mate" with each other and transmit their genes to the next generation. Commonly used fitness functions are the goodness of fit, root mean square error or leave-one-out (LOO) cross-validation coefficient in a regression model, accuracy or F-1 score in a classification model, and the free energy change or desired property value in a molecular optimization problem. Cross-over or recombination of two chromosomes leads to the exchange of genes between them, as shown in Fig. 6.1a. What the crossover procedure accomplishes is to break up the fittest group of features or substructures, and swap and recombine features or substructures between them, to create new groups in the next generation,

Table 6.1 Comparison of genetic algorithms used for feature selection and for molecular structure generation

GA application	Feature selection	Molecular structure generation
Goal	Find the best set of features for a QSAR model	Generate novel molecules or new molecular structures with desired properties
Search space	Space of molecular descriptors	Chemical space or space of molecular structures
Gene	A molecular descriptor or feature	A molecular substructure, substituent atom or functional group
Chromosome	A group of descriptors in a model	A group of substructures forming a molecule
Population	A collection of models	A set of molecules
Generation	A step in the feature selection process	A step in the molecular optimization process
Fitness function	Leave-one-out (LOO) cross-validation coefficient, accuracy, root mean square error, cross-entropy loss or goodness of fit	Free energy change, desired property value
Cross-over or recombination	Break up the fittest feature subsets or substructures, swap and recombine features between them, to create new feature subsets	Break up the fittest combination of substructures, swap and recombine substructures between them, to create new molecules
Mutation	Dropping or including a random descriptor in the model	Dropping or including a random substructure in the molecule

Fig. 6.1 (**a**) One-point crossover or recombination of chromosomes, and (**b**) point mutation operation, or flipping a random bit in the chromosome, in a genetic algorithm

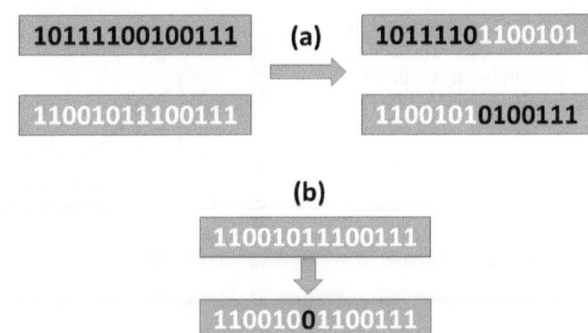

which then compete with each other and with the parents. The mutation operation consists of flipping a random bit in the chromosome, as shown in Fig. 6.1b, resulting in dropping $(1 \rightarrow 0)$ or including $(0 \rightarrow 1)$ a descriptor in the model or a substructure in the molecule. The recombination and mutation operations are applied independently to generate variance within the population, resulting in evolution of the population towards an optimized model or molecule.

While GAs are generally not explicitly formulated in network terms, the analysis in Fig. 1.7 should make clear that GAs and other feature selection methods are examples of growing and pruning the feature network. Likewise, molecular optimization is a process of growing and pruning the molecular graph, and generative chemistry involves exploration of chemical space, and growth of the chemical space network. Unlike GAs, that belong to the class of optimization methods known as evolutionary computation, all the other generative algorithms we will discuss in this chapter are ANNs of different architectures.

6.2 Back Propagation and Variational Auto Encoders

A feed-forward Artificial Neural Network (ANN) consists of layers of neurons, with each neuron i receiving inputs x_j from neurons j in the previous layer, performing a computation that is some function σ of the weighted sum of those inputs, with weights w_{ij}:

$$y_i = \sigma \left(\sum_j w_{ij} x_j + b \right) = \sigma(z_i) \qquad (6.1)$$

and feeding its output y_i to neurons in the subsequent layer, as shown in Fig. 6.2.

Hidden layers are layers between the input layer and the output layer. Deep neural networks (DNN) [4] are composed of many hidden layers. A wide neural network [5] is one that has a large number of neurons in each hidden layer. Here b is known

Fig. 6.2 The computation
performed by a single node
or neuron in an Artificial
Neural Network (ANN)

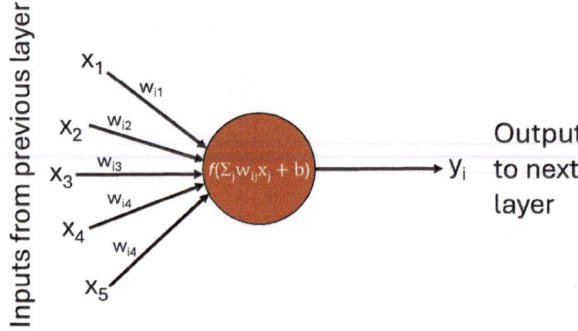

as a bias vector, and the function σ is called an activation function. Feed-forward
ANNs consisting of fully connected neurons with a nonlinear activation function are
also known as multi-layer perceptrons (MLP). The simplest activation function for
binary classification is the Heaviside step function:

$$H(z) = \begin{cases} 1 & \text{if } z \geq 0 \\ 0 & \text{if } z < 0 \end{cases}.$$ (6.2)

A smooth analytic alternative to the step function is the sigmoid function or the
standard logistic function, shown in Fig. 6.3a:

$$S(z) = \frac{1}{1 + e^{-z}}.$$ (6.3)

The softmax activation function, also known as the normalized exponential
function, is a multi-dimensional generalization of the logistic function:

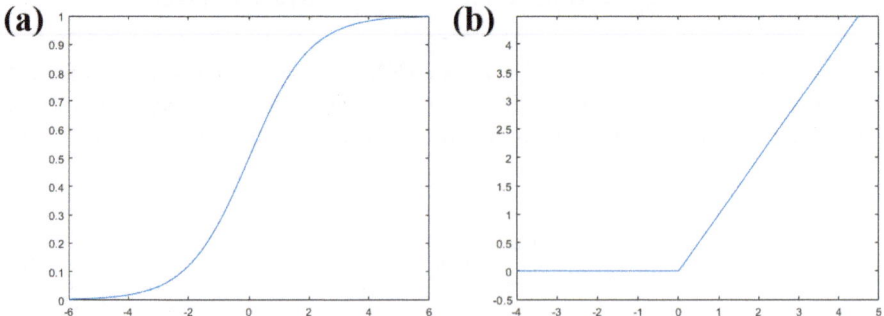

Fig. 6.3 a The sigmoid function or the standard logistic function, and **b** the rectified linear unit
(ReLU) activation function

$$\sigma(z_i) = \frac{\exp(z_i)}{\sum_{j=1}^{N} \exp(z_j)}. \tag{6.4}$$

where $\mathbf{z}$ is a vector in N dimensions: $\mathbf{z} \in \mathbb{R}^N$ and $\sigma(\mathbf{z}) \in (0,1)^N$. It maps a vector of real numbers into a probability distribution over the outputs, and is commonly used in multinomial logistic regression. The most popular nonlinear activation function in DNNs is the rectified linear unit (ReLU), shown in Fig. 6.3b:

$$\mathrm{ReLU(z)} = \max(z, 0). \tag{6.5}$$

Learning in supervised ANNs is accomplished through error back propagation, where the output value of the final, output layer is compared to the target value, and the discrepancy between the two (the error or loss L) is propagated back from the output layer to appropriately modify the weights and biases in previous layers. Popular choices for the loss function or cost function include sum of squared errors and cross entropy. First the gradient of the loss function with respect to the output is computed, and the chain rule is applied iteratively to compute the gradient of the loss with respect to the weights and biases of the previous layer:

$$\frac{\partial L}{\partial z} = \left(\frac{\partial L}{\partial y}\right)\left(\frac{\partial y}{\partial z}\right)$$

$$\frac{\partial L}{\partial w} = \left(\frac{\partial L}{\partial z}\right)\left(\frac{\partial z}{\partial w}\right) \tag{6.6}$$

$$\frac{\partial L}{\partial b} = \left(\frac{\partial L}{\partial z}\right)\left(\frac{\partial z}{\partial b}\right) = \frac{\partial L}{\partial z}$$

The computed gradients are then back propagated, to update the weights and biases of each previous layer using an optimization algorithm like gradient descent. The process is repeated until the ANN (hopefully) converges. ReLU results in faster learning in networks consisting of many layers (DNNs), in comparison to other activation functions, because the process of taking derivatives for error back propagation becomes really simple, and is driven to zero for $z < 0$.

An *auto-encoder*, as shown in Fig. 6.4, consists of two feed forward ANNs: the first with a bottleneck architecture, is known as the encoder, and is trained to map the data from the input layer into a low-dimensional latent vector. The second, the decoder, has an inverse bottleneck architecture, and maps the latent vector back into a high-dimensional chemical representation, so as to reconstruct the input data. The auto-encoder is trained to minimize the difference between the input chemical representation and the output of the decoder. Due to this challenge, the model is forced to learn an efficient encoding that represents the molecule as a small vector of fixed size, minimizing the loss between the input structures and the reconstructed ones. These learned embeddings are seen to cluster similar molecules close together in latent space. In a *variational autoencoder* (VAE), the molecule is represented as

Fig. 6.4 Schematic diagram
of an auto-encoder

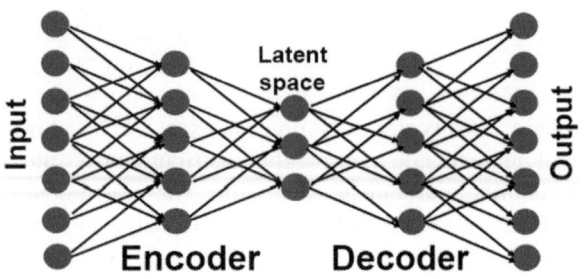

a continuous probability distribution in the latent space [6], which facilitates the computation of derivatives through the chain rule (Eq. 6.6) in the error back propagation step. The encoder and decoder are trained jointly to reconstruct the probability distribution of the training molecules. The VAE encoder thus performs the mapping $\mathbb{R}^N \to \mathbb{R}^M$ (M ≪ N), while the decoder maps $\mathbb{R}^M \to \mathbb{R}^N$. The latent vector encodes a compressed representation of molecular similarity in a space of low dimensionality, so that nearby points in the latent space of a VAE have similar chemical properties. This mapping from a discrete chemical space network to a continuous probability distribution in the latent space enables specific regions of latent space, for instance regions that contain ligands known to bind to a given receptor, to be sampled and decoded through the back propagation process, to generative novel molecules in a desired property range.

6.3 Graph Convolutional Networks

Graphs are natural representations of chemical data, with the atoms in a molecule forming the nodes of the molecular graph and the bonds constituting the edges. An ML algorithm that learns chemical data structures needs to preserve some invariances, such as invariances with respect to translations and rotations of the atomic coordinates, and to changes in the order of specification of the atoms. Graphs representations naturally preserve translational and rotational invariance, and eigenvalues of matrix representations of molecular graphs preserve permutational invariance. In graph neural networks, the molecular graphs as represented by adjacency matrices, and additional atomic, bond and molecular descriptors.

Graph Convolutional Networks (GCN) [7, 8] are a sub-class of *Convolutional Neural Networks* (CNN) [9], a kind of ANN originally designed for classifying 2D images, whose architecture was inspired by structure of the visual cortex of the mammalian brain. CNNs are built up of stacked convolutional, pooling and fully-connected or dense layers. A convolutional layer is a filter that determines the overlap of one function (the filter) as it is shifted over another function (the input) by computing the sum of the element wise product of the filter matrix and the window matrix, as shown in Fig. 6.5a and Table 6.2a–d. Each hidden layer in a CNN performs feature identification in a hierarchal manner, with the first few layers extracting low-level image representations such as edges and corners, and each successive layer transforming its input into a more abstract representation, and the final layers learning high-level representations. Convolutional layers are often followed by pooling layers such as max pooling, which extracts the most prominent feature in a patch of the image, as shown in Table 6.2e, or sum/average pooling, shown in Table 6.2f–g, which performs a coarse graining of features in the patch. Pooling layers generate translation invariant features that are robust to small shifts and distortions, and also reduce the dimension of the representation, thereby reducing the number of parameters to learn. CNNs have been widely employed in image processing, but also in other applications such as analysis of DNA and amino acid sequences.

In a GCN, schematically depicted in Fig. 6.5b and c, successive graph convolutional layers transmit each atom's attributes (atomic descriptors such as atom type, chirality, ring size, hybridization, aromaticity, donor or acceptor character) to its bonded neighbors, and likewise the bond attributes (such as bond order, aromaticity

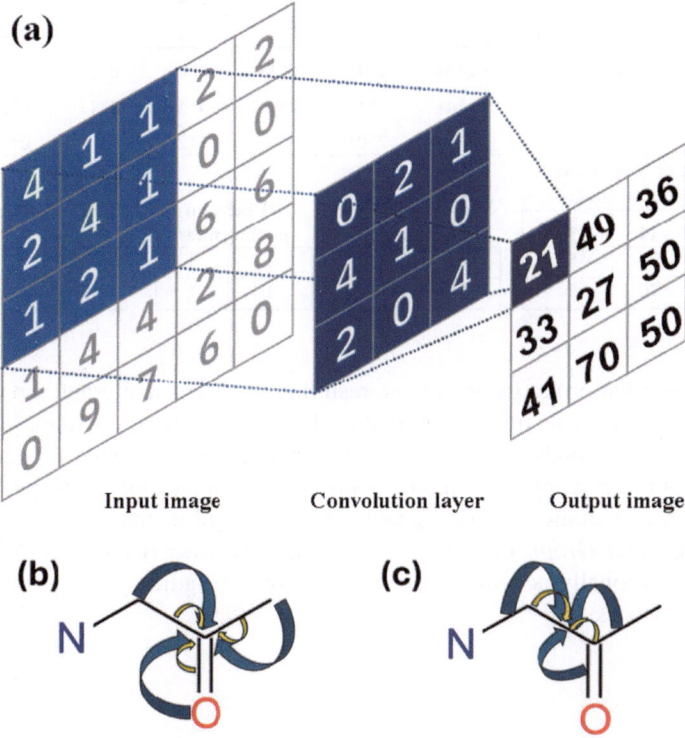

Fig. 6.5 (a) Schematic diagram of the operation of a convolutional neural network (CNN). (b) Schematic diagram showing how information is passed to a node/atom and (c) to an edge/bond in a graph convolutional neural network (GCN) from the adjoining atoms and bonds of the molecular graph

Table 6.2 (**a**) Input to a convolutional neural network (CNN), and (**b**) a 3×3 convolution layer, (**c**) application of the convolutional filter to the input, its result (**d**), (**e**) the result after 2×2 max pooling, (**f**) a 2×2 sum filter; dividing the output by 4 (the number of pixels in the filter) results in average pooling (**g**)

(a)

Input Image				
4	1	1	2	2
2	4	1	0	0
1	2	1	6	6
1	4	4	2	8
0	9	7	6	0

(b)

Convolution Filter		
0	2	1
4	1	0
2	0	4

(c) **Application of the Filter to the Image**

4*0+	1*2+	1*1+	1*0+	1*2+	2*1+	1*0+	2*2+	2*1+
2*4+	4*1+	1*0+	4*4+	1*1+	0*0+	1*4+	0*1+	0*0+
1*2+	2*0+	1*4	2*2+	1*0+	6*4	1*2+	6*0+	6*4
2*0+	4*2+	1*1+	4*0+	1*2+	0*1+	1*0+	0*2+	0*1+
1*4+	2*1+	1*0+	2*4+	1*1+	6*0+	1*4+	6*1+	6*0+
1*2+	4*0+	4*4	4*2+	4*0+	2*4	4*2+	2*0+	8*4
1*0+	2*2+	1*1+	2*0+	1*2+	6*1+	1*0+	6*2+	6*1+
1*4+	4*1+	4*0+	4*4+	4*1+	2*0+	4*4+	2*1+	8*0+
0*2+	9*0+	7*4	9*2+	7*0+	6*4	7*2+	6*0+	0*4

(**d**) **Result of application of Filter to Image**

21	49	36
33	27	50
41	70	50

(**e**) After 2x2 max pooling

49	50
70	70

(**f**) 2x2 sum pooling filter

1	1
1	1

(**g**) After 2x2 average pooling of (d)

32.5	40.5
42.75	49.25

and ring information) to its neighbors, resulting in an updated graph representation with coarse-grained features, such that each atom and bond contains information about atoms and bonds in its local neighborhood. This sequential update scheme provides for information flow among atoms, bonds and the molecular state. Stacking more graph convolutional layers enables longer-range interactions to be incorporated. In *Crystal Graph Convolutional Neural Networks* (CGCNN) [10–12], one needs to additionally account for invariances due to lattice periodicity and space group symmetries.

6.4 Generative Adversarial Networks and Reinforcement Learning

AI algorithms employing *Generative Adversarial Networks* (GAN) [13] have of late been used extensively to create art, music, poetry, stories and essays rivaling humans in creativity. GANs based on a molecular input representation such as SMILES strings, fingerprints, or molecular graphs, have also been used for inverse design, a long standing dream of the cheminformatics/materials informatics community, which aims to generate novel molecules designed to have a desired set of properties for specific applications. GANs were inspired by game theory, and consist of two DNNs—a *Generator* and a *Discriminator*, that compete against each other in a machine learning game, as shown in Fig. 6.6. The generator is trained to generate new molecules matching the distribution of the input molecules, while the discriminator's goal is to distinguish the input molecules from those generated by the generator. In the process of competing in this adversarial game, both generator and discriminator improve their performance at their respective tasks, with the generator getting better at producing molecules possessing the desired characteristics.

GANs have been fruitfully employed in several generative chemistry applications. For instance, Kadurin *et al.* developed DruGAN [14], a GAN with an autoencoder architecture designed for *de novo* generation of novel molecules with anticancer properties. Méndez-Lucio *et al.* [15] conditioned a GAN with transcriptomic data to generate inhibitor-like ligands only using the gene expression signatures, without the need for molecules with labelled activities, thereby guiding molecular generation towards specific desirable areas of chemical space.

Although GANs work well with continuous variables, gradients for back-propagation are undefined when the output is discrete, like molecular structures. So GANs are often combined with *Reinforcement Learning* (RL) [16, 17]. RL avoids the use of brute-force to examine all possible solutions to a problem, but instead learns actions by trying them out and making mistakes, a strategy that has enabled such

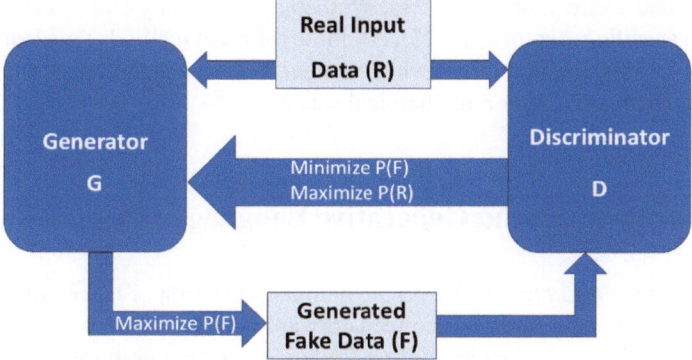

Fig. 6.6 Schematic diagram of a generative adversarial network (GAN)

algorithms to beat the best human champion in Go, a game with a vast solution space. RL thus represents a promising method for exploring the immense chemical space in drug and materials discovery. In RL, an agent discovers the actions that yield the maximum reward by interacting with a dynamic environment and learning an optimal policy by trial and error. The policy actions are chemical transformations or modifications of the molecular structure, such as addition or removal of atoms, functional groups, rings and/or bonds. After the agent selects an action, a reward function evaluates it against a target property, such as synthetic accessibility or quantitative estimate of drug-likeness (QED), and provides feedback. The agent thus learns the actions that maximize the cumulative reward [18].

Supplementing GANs with RL serves to balance the demands of exploitation of a structure–activity relationship (by generating molecules similar to the initial distribution) and exploration of new chemical space (by the RL, which attempts to shift the distribution of generated molecules in a direction that optimizes the desired properties). In algorithms such as Objective-Reinforced Generative Adversarial Network for Inverse-design Chemistry (ORGANIC) [19], Reinforcement Learning for Structural Evolution (ReLeaSE) [20], and Reinforced Adversarial Neural Computer (RANC) [21], both generator and discriminator are trained directly on SMILES representations of molecules from a large database, with every new generated molecule earning either a numerical reward or a penalty. The generator is trained to generate molecular structures, while the discriminator rewards or penalizes the structures generated based on drug-likeliness, ease of synthesis or other target properties. The model thus learns to relate similarities between SMILES strings to similarities between molecular properties, and thence to optimize these properties for the generated molecules. In ReLeaSE, the models are trained in two stages, with the generator and discriminator being trained separately at first in a supervised manner, and then jointly using RL to optimize the target properties. Both RANC and ORGANIC combine GAN with RL, and are designed to recognize significant molecular patterns and structural scaffolds, and to generate novel and easy to synthesize molecules for specific targets.

Graph-based generative RL models combining GCNs with RL have also been used for multi-objective optimization, to design novel molecules with multiple constraints such as potency, safety, drug-likeness, synthetic accessibility and selective binding potency to specific receptors [22–25]. All these different methods thus learn a representation of molecular similarity, which is then used to expand the chemical space network in the desired region of chemical space.

6.5   Transformers and Generative Language Models

In a *recurrent neural network* (RNN) the output of a neuron or layer can also be one of its own inputs, in contrast to the feed-forward neural networks we have considered until now. Thus the topology of an RNN is a directed graph with one or more cycles (self-loops), enabling it to perform recursion. This gives rise to persistence or "memory" in the network. RNNs process sequential input data, such as strings of text, with no predetermined limit on size (unlike CNNs), and have been used with molecular SMILES representations in generative models to grow a large chemical space network from a very small sample thereof. For instance, Arús-Pous *et al.* [26] used the GDB-13 dataset that consists of 975 million small organic molecules with up to 13 C, N, O, S and Cl atoms passing chemical stability and synthetic feasibility filters. A model trained on only 0.1% of the database (1 million molecules) was able to grow the network to reproduce nearly 70% of the database.

However, RNNs are slow to train and suffer from exploding gradient and vanishing gradient problems; their memory is not strong when parsing long sequences, due to the domino-like effect of recurrence on back-propagation when gradients are computed through the chain-rule (6.6). Any change in the weights iteratively affects the flow of information from one layer to the next; so the gradients can rapidly become either very large or very small. The vanishing gradient problem can happen when the gradients that update the weights become smaller and smaller and eventually become insignificant or "vanish" as they are back-propagated from the output layer to earlier layers. At this point the weights can no longer be updated, and the network has stopped learning. As a result, the memory is weak for far-away words or tokens in long text sequences, and the RNN has difficulty processing long text passages. Conversely, when the gradients are too large, the model weights continue to grow, leading to instability—the exploding gradient problem. Consequently, RNNs have largely been supplanted by the attention mechanism in recent years.

The *transformer* preprint of Vaswani *et al.* [27], that introduced the concept of self-attention, was perhaps the most transformative AI paper of recent times. Transformers were first applied in the realm of natural language processing for machine translation and computer vision, but are now ubiquitous in all kinds of applications, and form the basis of modern Large Language Models (LLMs). The self-attention mechanism applied at each step of parsing the input string captures the relationships between different words in a sentence or between different characters in a SMILES string, regardless of their relative position, thereby enabling the model to "attend to" the relevant parts of the input string. At any position, the self-attention mechanism examines the previous tokens (words or characters) in the input string and scores them in order of importance for predicting the next token. This ensures the encoding of long-range correlations between tokens in the sequence. Each attention layer employs three parameters: a query vector ($\mathbf{Q}$) and keys vector ($\mathbf{K}$) of dimension d_k, and a values vector ($\mathbf{V}$) of dimension d_v:

$$\text{Attention}\,(\mathbf{Q}, \mathbf{K}, \mathbf{V}) \;=\; \text{Softmax}\!\left(\frac{\mathbf{Q}.\mathbf{K}^{\mathrm{T}}}{\sqrt{d_k}}\right)\mathbf{V} \tag{6.7}$$

This "scaled dot product attention" computes the inner products or dot products of the query with all keys, before scaling by $\sqrt{d_k}$ and applying the softmax function, to give the weights on the values $\mathbf{V}$, as shown in Fig. 6.7a. As in natural language, the influence of an atom on molecular properties depends upon its position and its neighbors. Position-dependent sine and cosine functions are thus added to the input embedding in order to encode the order of tokens in a sequence. Typically,

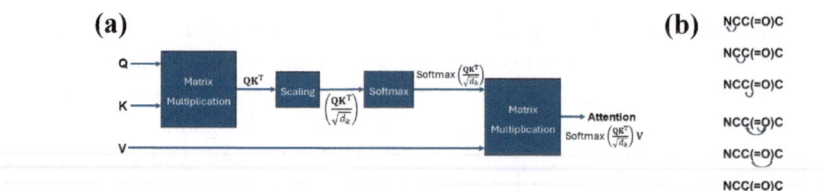

Fig. 6.7 (a) Scaled dot-product attention in a transformer. (b) The previous tokens that must be paid attention to at each position of the SMILES string NCC(=O)C in order to parse the string correctly

multiple attention layers are applied in parallel, and their outputs concatenated, to constitute a multi-head attention block, enabling the transformer to pay attention to information from different positions in each layer. For instance, Fig. 6.7b shows the previous tokens that must be paid attention to at each position of the SMILES representation NCC(=O)C of the molecule depicted in Fig. 1.2, in order to parse the string correctly. Typically a transformer consists of several multi-head attention blocks and fully connected feed-forward layers, and can have billions of trainable parameters. Since the attention vectors are independent of one another, transformers can process the entire input stream in parallel, and can be trained fast using GPUs.

The multi-head self-attention mechanism may be considered as an on-the-fly construction of a co-occurrence network between tokens. Each node of this network is a token in the input string, and the edges incident upon any node identify the important previous tokens in the string that need to be attended to at that position. The attention mechanism can also be related to the idea of Lagrangian duality [28] referred to in Sect. 2.3.

There have been may applications of transformers in generative chemistry, trained on molecules in SMILES representation. Because large databases of labelled chemistry data (molecules with experimentally validated activities) are uncommon, these applications generally employ a strategy of pre-training on large datasets for simple end points, essentially to teach the model to generate valid SMILES strings, followed by fine-tuning with the limited available data on the desired property. This strategy is also known as *transfer learning*. Rather than concentrating on any single property prediction task, pre-trained transformers aim to generate a context-rich generalizable embedding or representation of molecules in latent space. This pre-training is often done in an unsupervised manner, with unlabeled data. For instance, a generative pre-trained molecular transformer model, MolGPT [29], with around 6 million parameters, was developed for *de novo* generation of novel drug-like molecules. The model was first pre-trained on a masked self-attention task, namely to predict the next token in a SMILES string, hiding all SMILES tokens to the right. Then the model, having learned the SMILES syntax during pre-training, was fine-tuned with labeled data, to predict molecular properties. The generated molecules showed high validity (the fraction of generated structures that were valid SMILES strings), uniqueness (the fraction of valid generated molecules that were unique), novelty (the fraction of valid unique molecules generated outside the training set), and internal diversity.

They could be further restricted to generate molecules within specified ranges of properties such as logP (the logarithm of the octanol–water partition coefficient), synthetic accessibility, topological polar surface area (a measure of permeability across cell membranes), drug-likeness or a combination thereof.

Transformers have also been employed for *de novo* design of proteins to catalyze chemical reactions. ProteinGAN [30] incorporates a self-attention mechanism into a GAN to learn the evolutionary relationships of proteins directly the amino acid sequences, and then generate novel proteins with specific functions, including sequence motifs not represented in the training set.

Another novel idea was to treat *de novo* drug design as a translation task from the protein sequence space of the target (in amino acid language) to the chemical space of drug molecules (represented in SMILES language). Solely from the amino acid sequence, without knowing the 3D structure of the protein or having libraries of active drug molecules to train on, a transformer model was able to generate new molecules predicted to bind the target protein [31].

Another transformer language model, ProtGPT2 [32], with 738 million parameters, was trained on ~ 50 million natural protein sequences, and was able to generate novel protein families, exploring new regions of protein space.

Transformer-based language models have been employed to predict polymer mechanical and thermal properties, as represented by the SMILES strings of their monomer repeat units. TransPolymer [33] was first pre-trained on a large, unlabeled polymer dataset, and thereafter fine-tuned on datasets of polymer properties, such as electrical conductivity, band gap, dielectric constant, refractive index, ionization energy, electron affinity, crystallization tendency, and organic photovoltaic power conversion efficiency. TransPolymer was able to outperform other ML models trained with chemical fingerprints on most of these tasks.

A variant of the original transformer idea is Devlin's deep Bidirectional Transformer (BERT) [34], where the self-attention mechanism is applied in both directions—left to right as well as right to left. BERT is pre-trained using several encoder blocks in sequence, without any decoder blocks, on a masked language learning task, where some fraction of the input tokens, selected at random, are masked or corrupted, and the model is trained to reproduce the original string. BERT has been employed in many chemical and biological language models [35–41], such as:

- SMILES-BERT [35], a semi-supervised attention-based model, consisting of six 4-head transformer encoder layers, that was pre-trained on the ZINC database (a large unlabeled collection of over 35 million molecules) through masked SMILES recovery, and then fine-tuned to predict various molecular properties.
- ChemBERTa-2 [36], a chemical foundation model for prediction of molecular properties, that employs a BERT-like transformer with 5–46 million parameters, and is pre-trained on up to 77 million SMILES strings from PubChem, using masked SMILES recovery and multi-task regression. The pre-trained models were fine-tuned and tested on properties such as blood–brain penetrability, clinical toxicity, solubility, and HIV inhibition.

- polyBERT [37], a bidirectional transformer-based chemical language model for polymers, pre-trained on 100 million virtual polymers (generated by decomposing 13,766 known polymers into unique chemical fragments, and then recombining them) to predict masked tokens using the surrounding tokens, and then fine-tuned to predict 29 polymer properties.
- BatteryBERT [38], a language model for battery research, pre-trained on a battery-related research papers, which yielded 1870 million word tokens (compared to the 3300 million in BERT). The pre-trained model was fine-tuned for classifying battery-related research papers, and battery device components into anode, cathode and electrolyte materials.
- BioBERT (Bidirectional Encoder Representations from Transformers for Biomedical Text Mining) [39], a BERT-based language model for biomedical text mining pre-trained on biomedical literature. BioBERT outperformed other models on biomedical question answering, named entity recognition and biomedical relationship extraction.
- DNABERT [40], a bidirectional transformer model pre-trained on nucleotide sequences, to decode the language of non-coding DNA in the genome. It has 12 transformer layers and 768 hidden units, with 12 attention heads in each layer. After fine-tuning the pre-trained model with small task-specific datasets, state-of-the-art performance was achieved on prediction of promoters, splice sites and transcription factor binding sites. Visualization of importance and semantic relationship within the sequences provide interpretability.
- ProteinBERT [41], a BERT-based language model for learning protein sequence and function, pre-trained on over 100 million protein sequences and Gene Ontology annotations from the UniProtKB and UniRef90 databases. Pre-training involved recovering both the protein sequence and the known Gene Ontology annotations. The architecture of ProteinBERT consists of both local and global representations, allowing it to process protein sequences of any length and generalize to unseen sequence lengths. ProteinBERT achieved state-of-the-art performance, with a smaller and faster model, on different tasks such as protein structure prediction and prediction of post-translational modifications.

6.6 Why Does Over-Parametrization Work?

At this point we need to address the question of why massively over-parametrized networks such as DNNs, and in particular LLMs, with millions or billions of parameters (1.76 trillion in the case of GPT4) are able to generalize outside the training set. An over-parametrized model is one where the number of adjustable parameters far exceeds the number of training samples. As we have seen, over-parametrized models should be able to fit the training data perfectly by just "memorizing" the training data, including the noise therein, achieving zero training error, but according to traditional wisdom, we would expect poor performance on new, unseen data due to overfitting, resulting in poor generalization ability of the models. But surprisingly, massively over-parametrized neural networks seem to generalize reasonably well within the applicability domain.

Generalization error in traditional machine learning follows a U-shape curve: as more parameters are added, the training error continuously decreases, but the test error starts to increase past a certain threshold, which separates the under-fitting regime from the overfitting regime. But for deep learning models, Belkin *et al.* [42] proposed and empirically observed the double-U-shaped curve shown in Fig. 6.8.

Actually, the number of parameters does not indicate the true complexity of DNNs, as only a subset of the network parameters have an impact on the model performance [42–44]. A randomly initialized, dense, feed-forward DNN contains several sub-networks, only a few of which can achieve optimal performance when trained in isolation. This can be seen by pruning the network, i.e. removing edges with negligible weights; this can be done without degrading the model performance. A large number of parameters is thus not needed for representing the final, optimized solution, but a large parameter space provides a large pool of initialization configurations of many small sub-networks during training. Thus the number of parameters is not

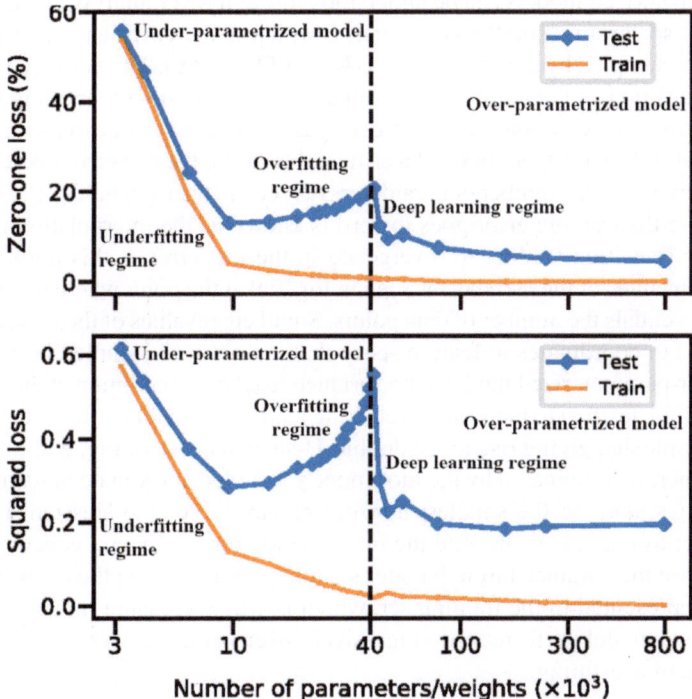

Fig. 6.8 Double descent risk curve for fully connected neural network on the MNIST dataset. Training and test risks of network with a single layer of H hidden units, learned on a subset of the MNIST dataset ($n = 4 \times 10^3$, $d = 784$, $K = 10$ classes). The number of parameters is $(d + 1)H + (H + 1)K$. Adapted with permission from Ref. [42]. © 2019, National Academy of Sciences. All rights reserved

correlated with model overfitting in such networks, and thus such networks are not necessarily overfitted.

A much better metric is the intrinsic dimension, and many problems have much smaller intrinsic dimensions than the number of parameters. In an over-parametrized neural network, the parameters form a high-dimensional manifold upon which learning takes place. With a smooth manifold, there are more predictive gradients, which allows for larger learning rates, and this helps in optimization.

Although the parameter space is huge, local optima are almost always saddle-points, and thus the possibility of the optimization getting stuck there is low, as there are always coordinates along which gradient descent can continue. But because the parameter space is high dimensional, the network does not have to explore all these dimensions to learn efficiently. Exploring only a sub-network of the parameter landscape can often provide a reasonable solution, and thus the true complexity of the model or the intrinsic dimension is much lower than the dimensionality of the parameter space. In over-parameterized networks, gradient descent converges to the global optimum at a linear convergence rate.

Rocks and Mehta [45] showed how the behavior of the generalization error in over-parameterized models can be understood through RMT analysis, in terms of the eigenvalue spectrum of the Hessian matrix, which is the covariance matrix among the features sampled by the training data (K_{train} of Eq. 2.59). The network undergoes a phase transition as the number of parameters is increased—the training error first decreases to zero while the test error diverges after an initial decrease (as seen in the left half of Fig. 6.8), but beyond a critical threshold, the test error decreases with further increase in the number of parameters (as seen in the right half of Fig. 6.8). The point where the training error goes to zero is known as the interpolation threshold. The phase transition leading to divergence in the test error at this point is due to small eigenvalues of the Hessian, as seen in Fig. 6.9 at the point where the number of parameters equals the number of data points. Small eigenvalues of the Hessian matrix correspond to coordinates in feature space that are not well sampled in the training set. In over-parameterized models, the variance reaches a maximum at the interpolation threshold, leading to increased overfitting tendency, but thereafter decreases with model complexity, giving rise to the double-U-shaped curve of Fig. 6.8. Model variance is generally dominated by the most poorly sampled coordinate in feature space, which corresponds to the smallest nonzero eigenvalue of the Hessian. Increasing the number of parameters beyond the interpolation threshold can reduce overfitting and decrease the variance through better sampling of data along the coordinates that are well represented in the training set, which is why increasing model complexity can improve model performance in massively over-parametrized networks without giving rise to overfitting.

Fig. 6.9 Analytic solution
for the minimum eigenvalue
of the Hessian matrix.
Reprinted under CC-BY-4.0
license from [45]. © 2022,
The Author(s)

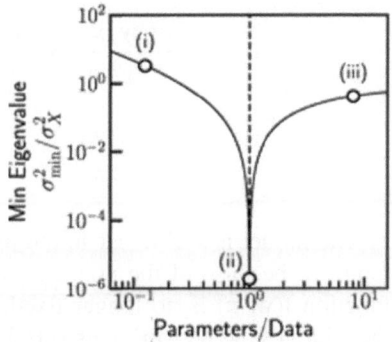

6.7   Infinitely Wide Networks and Neural Tangent Kernels

Now let us turn to wide networks. Specifically, we will investigate the limit wherein
the number of neurons in a layer approaches infinity. Consider a fully connected
ANN, consisting of hidden layers, each containing $n_0, n_1, n_2, \ldots n_{\lambda-1}$ neurons, and
let us collect all the p parameters (weights and biases) of the network into a weight
vector:

$$\mathbf{w} = \left(w_0, \ w_1, \ w_2, \ w_3, \ldots w_p \right) \tag{6.8}$$

which is randomized initially and then trained with gradient descent. Let us examine
the evolution of any hidden layer j (with n_j neurons) of a wide neural network under
gradient descent. Consider a scalar network function $f(\mathbf{w}, \mathbf{x})$ which maps the inputs
$\mathbf{x}$ to the outputs:

$$y = f(\mathbf{w}, \mathbf{x}) = \frac{1}{\sqrt{n_j}} \sum_{d=1}^{n_j} w_{j,d} \sigma \left(\mathbf{w}_{j-1,d}^T \mathbf{x} + \mathbf{b} \right), \tag{6.9}$$

where $w_{j-1,d}$ contains the weights that relate the input $\mathbf{x}$ to the dth node of hidden
layer j and $w_{j,d}$ is the weight that relates the dth node to the output. The factor $1/\sqrt{n_j}$
ensures consistent asymptotic behavior as the widths of the hidden layers $n_0, n_1, n_2,$
$\ldots n_{\lambda-1} \to \infty$. As in Sects. 2.2 and 2.3, we can perform gradient descent using a
squared loss function:

$$L(\mathbf{w}) = \frac{1}{2} \sum_i \left\{ f(\mathbf{w}, x_i) - y_i \right\}^2, \tag{6.10}$$

where the summation is over the training pairs (x_i, y_i).

$$\nabla_w L\left(\mathbf{w}^{(t)} \right) = \sum_i \left\{ f(\mathbf{w}, x_i) - y_i \right\} \nabla_w f(\mathbf{w}, x_i) \tag{6.11}$$

$$\frac{\partial f(\mathbf{w}, \mathbf{x})}{\partial w_{j-1,d}} = \frac{1}{\sqrt{n_j}} \sum_{d=1}^{n_j} \sigma\left(\mathbf{w}_{j-1,d}^T \mathbf{x} + \mathbf{b}\right).w_{j,d}\mathbf{x} \qquad (6.12)$$

$$\frac{\partial f(\mathbf{w}, \mathbf{x})}{\partial w_{j,d}} = \frac{1}{\sqrt{n_j}} \sum_{d=1}^{n_j} \sigma\left(\mathbf{w}_{j-1,d}^T \mathbf{x} + \mathbf{b}\right) \qquad (6.13)$$

The model is still linear in the model parameters $w_{j,d}$, but it is nonlinear in the inputs $\mathbf{x}$, because of the nonlinear activations σ. Furthermore, the gradient of the function $f(\mathbf{w}, \mathbf{x})$ is no longer fixed, but changes during the training. However, as the width of the network gets very large, it is empirically observed that the weights are almost static, and the network becomes approximately linear with respect to the parameters $\mathbf{w}$. This is called "lazy training." In the limit of infinite width i.e. as $n_j \to \infty$, the gradient update is distributed over the infinitely many weights in the layer. Each parameter changes negligibly during training, but collectively they contribute to a finite change in the output. So the output f can thus be expressed as a first-order Taylor expansion in the weights $\mathbf{w}$:

$$f(\mathbf{w}, \mathbf{x}) \approx f\left(\mathbf{w}^{(0)}, \mathbf{x}\right) + \nabla_w f\left(\mathbf{w}^{(0)}, \mathbf{x}\right)^T.\left(\mathbf{w} - \mathbf{w}^{(0)}\right), \qquad (6.14)$$

where $f(\mathbf{w}^{(0)}, \mathbf{x})$ is the output for the given input vector $\mathbf{x}$ and the original weight vector $\mathbf{w}^{(0)}$, $f(\mathbf{w}, \mathbf{x})$ is the corresponding output for the updated weight vector $\mathbf{w}$, and $\nabla_w f(\mathbf{w}^{(0)}, \mathbf{x})$ is the tangent vector at initialization. Thus each weight update is approximately given by the linear term in Eq. (6.14), and we will show below that as the number of nodes in the layer $n_j \to \infty$, the neural network evolves as a linear model under gradient descent.

We can think of $\nabla_w f(\mathbf{w}, \mathbf{x})$, the gradient of the output with respect to the model parameters, as a potentially nonlinear transformation of the inputs:

$$\varphi(\mathbf{x}) = \nabla_w f(\mathbf{w}, \mathbf{x}). \qquad (6.15)$$

We can then use this function to define a kernel, the *Neural Tangent Kernel* (NTK [46–48]), as in Eq. (2.46):

$$K(\mathbf{x}, \mathbf{x}') = \varphi(\mathbf{x})^T \varphi(\mathbf{x}') = \sum_p \frac{\partial f(x)}{\partial w_p}.\frac{\partial f(x')}{\partial w_p}.$$
$$= \nabla_w f(\mathbf{w}, \mathbf{x})^T.\nabla_w f(\mathbf{w}, \mathbf{x}'). \qquad (6.16)$$

This function is symmetric, positive semi-definite, and thus a valid kernel function. As we have seen in Sect. 2.3, a kernel function describes a similarity measure between two inputs, here $\mathbf{x}$ and $\mathbf{x}'$. The NTK describes how updating the model parameters on one molecule $\mathbf{x}$ affects the predictions for another molecule $\mathbf{x}'$, and it provides a link between ANN and kernel methods. For finite width networks, the NTK depends on the specific random choice of the parameters, but as $n_j \rightarrow \infty$, the kernel is constant; it does not depend on the initialization. It has been proven that a wide ANN, with a large number of neurons per layer, converges to a global minimum during training.

NTK describes the evolution of a DNN during training by gradient descent. The random initialization of the network parameters induces a distribution over f(x), which can be visualized as an ensemble of networks constructed from the same initial distribution f(x), and trained each using the same procedure. When the network is first initialized, the ensemble is a zero-mean maximum entropy distribution with no structure other than its mean and covariance, as a consequence of the central limit theorem, and the kernel is random. In general, the kernel depends upon the parameters, and varies during training, but in the infinite-width limit, i.e. as $n_j \rightarrow \infty$, it becomes constant, converging to a deterministic limit, and remaining unchanged during training:

$$K(\mathbf{x}, \mathbf{x}') \rightarrow K_\infty(\mathbf{x}, \mathbf{x}') \text{ as } n_j \rightarrow \infty. \tag{6.17}$$

When the parameters of the ANN are optimized with gradient descent during training, the network function f follows the kernel gradient of the cost functional with respect to the NTK; the ensemble evolves according to the neural tangent kernel.

In the limit when the learning rate goes to zero, $\eta \rightarrow 0$, we have from Eq. (2.24):

$$\frac{dw^{(t)}}{dt} = -\nabla_w L(w^{(t)}). \tag{6.18}$$

This is known as gradient flow dynamics. Defining U as the deviation of the predicted function value f($\mathbf{w}$, $\mathbf{x}$) from the actual value of y:

$$U = f(\mathbf{w}, \mathbf{x}) - y, \tag{6.19}$$

and using the squared loss function L(w), we have:

$$\nabla_w L(w) = \nabla_w f(\mathbf{w}, \mathbf{x})(f(\mathbf{w}, \mathbf{x}) - y) = \nabla_w f(\mathbf{w}, \mathbf{x}) U. \tag{6.20}$$

From Eqs. (6.18) and (6.19), we have:

$$\frac{dw^{(t)}}{dt} = -\nabla_w f(\mathbf{w}, \mathbf{x}) U \tag{6.21}$$

for the dynamics of the weights, and for the dynamics of the output:

$$\frac{df(w^{(t)})}{dt} = -\nabla_w f(\mathbf{w}, \mathbf{x})^T \frac{dw^{(t)}}{dt} = -\nabla_w f(\mathbf{w}, \mathbf{x})^T . \nabla_w f(\mathbf{w}, \mathbf{x}) U, \tag{6.22}$$

where we have used the chain rule and Eq. (6.21). Or using Eq. (6.16):

$$\frac{dU}{dt} \approx -K\big(w^{(0)}\big)U. \tag{6.23}$$

Thus wide networks approximate a linear model, and in the limit of infinite width, the network function follows a linear differential equation during training, leading to better interpretability and generalization ability. This differential equation has the solution:

$$U(t) = U(0)\exp\big\{-K\big(w^{(0)}\big)t\big\}. \tag{6.24}$$

When $t \rightarrow \infty$, $U(t) \rightarrow 0$, which means $f(w) \rightarrow y$, or zero training error.

Over-parametrized networks have the property that the Jacobian matrix remains positive definite (not just positive semi-definite) for all iterations, which implies that the smallest eigenvalue is non-zero. The corresponding eigenfunctions are the kernel principal components, and the convergence in Eq. (6.17) is thus fastest along the largest principal components of the input data vectors with respect to the NTK.

DNNs are intricately related to the Gaussian processes (GP) we discussed in Sect. 2.3. When an infinitely wide ANN is initialized with random weights, it is equivalent to a Gaussian process. The equivalence between GPs and deep, infinitely wide neural networks follows from the central limit theorem, and allows us to treat inference in DNNs as a kernel regression by evaluating the corresponding GP. Consider a fully-connected DNN with independent and identically distributed initial random weights and biases, each with zero mean, and weight variance scaled inversely proportional to the layer width. Since the output of each node in any hidden layer is a sum of independent and identically distributed terms, it follows that in the limit of infinite width ($n_j \rightarrow \infty$), each output will be Gaussian distributed, as a consequence of the central limit theorem, and thus the final output will have a joint multivariate Gaussian distribution, namely a Gaussian process. Lee *et al.* [49] have shown that the best generalizing networks are consistently the widest. Furthermore, the use of a GP prior means that every test data point has an associated uncertainty estimate, which is strongly correlated with the prediction error of the network [49].

6.8   Extensions and Future Directions

Perhaps the greatest problem with deep learning at present is its lack of interpretability. Ideally we would like to not just make accurate predictions, but also know why a model is making that particular prediction. We have seen how infinitely wide networks, by virtue of the fact that they approximate a linear model, lead to better interpretability. Kolmogorov-Arnold Networks (KAN [50]) represent another effort in this direction. In contrast to MLPs whose nodes have fixed activation functions, with learnable weights on their connections or edges, KANs have learnable activation functions on their edges, as shown in Fig. 6.10.

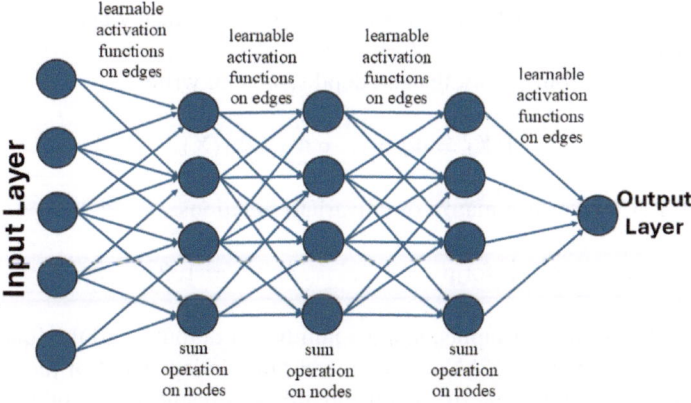

Fig. 6.10 Schematic diagram of a Kolmogorov-Arnold networks (KAN)

KANs are based on a generalization of Kolmogorov and Arnold's representation theorem [51] to networks of arbitrary widths and depths. According to this theorem, any multivariate function can be written using univariate functions and the operation of summation. Thus learning a high-dimensional function reduces to learning several univariate functions. The original formulation had only two-layer nonlinearities, with m neurons in the input layer and only a small number n of neurons in the hidden layer:

$$f(\mathbf{X}) = \sum_{k=1}^{n} \Phi_k \left(\sum_{j=1}^{m} \varphi_{kj}(x_j) \right) \tag{6.25}$$

where $\mathbf{X^i} = (x_1{}^i, x_2{}^i, \dots x_m{}^i)$ is the set of input descriptors and y_i the corresponding output for the ith molecule,

$$\varphi_{kj}(x_j) = \sum_{p=1}^{r} H_{kjp} \chi_p(x_j) \tag{6.26}$$

is the inner function, a linear combination of the r univariate basis functions χ_p with $m \cdot n \cdot r$ learnable coefficients H_{kjp}, $\Sigma_j \varphi_{kj}$ is the input to the kth neuron in the hidden layer, and

$$\Phi_k(z) = \sum_{q=1}^{s} G_{kq} \psi_q(z) \tag{6.27}$$

is the outer function, a linear combination of the s univariate basis functions ψ_q with $n \cdot s$ learnable coefficients G_{kq}. The output $f(\mathbf{X^i})$ is the sum of inputs from the n neurons in the hidden layer. The problem then reduces to finding the parameters H_{kjp}

and G_{kq} given the set of data $\{\mathbf{X}, \mathbf{y}\}$, such that the discrepancy between f($\mathbf{X}$) and $\mathbf{y}$ is minimized.

Generalizing to arbitrary widths and depths, we can write:

$$f(\mathbf{X}) = \Phi_n \circ \ldots \circ \Phi_2 \circ \Phi_1(\mathbf{X}) \tag{6.28}$$

where each KAN layer is a matrix of univariate functions:

$$\Phi_i = \{\phi_{q,p}\}, \text{ for p } = 1, 2, \ldots, n_{in}, q = 1, 2 \ldots, n_{out}, \tag{6.29}$$

where n_{in} is the number of inputs, n_{out} the number of outputs, and the functions $\phi_{q,p}$ have trainable parameters. KANs have an error rate independent of the dimension, thus circumventing the curse of dimensionality and resulting in greater accuracy, even with networks of smaller size, and thus have potentially greater interpretability, but their major disadvantage at present is the slow training.

Another limitation of methods based on matrices and networks with pairwise connections between nodes is that these structures cannot effectively capture higher-order relationships. For instance, the contribution of an atom to a molecular property is often not simply additive, but depends upon several other atoms in its neighborhood. Likewise, in feature networks, it is not just pair-wise correlations, but also higher-order correlations that are important. One then needs to deal with simplices and their associated tensor representations [52]. But that takes us beyond the scope of this book, which is concerned with molecular networks.

In this chapter, we have seen how deep learning and generative methods are being used for the design of novel materials, and how similarity kernels and chemical space representations are key to these advances. The last few years have seen an explosion of activity in the area of AI and deep learning in particular. Many of these advances have been exploited for drug and materials design, and many more applications are sure to follow. Certainly some of these directions might turn out to be less promising than initially believed and there might be other potential breakthroughs that have been overlooked in this book, but it has been my intention here to give the reader an overview of some of the most interesting developments in this area and some view of the what the future might possibly hold. I will conclude in the next chapter with some thoughts on the role and scope of creativity in molecular design.

References

1. V. Rasic, LeadDelta (2024). https://leaddelta.com/networking-quotes/
2. J. Holland, *Adaptation in Natural and Artificial Systems* (Addison Wesley Longman, Reading, 1989)
3. N. Sukumar, G. Prabhu, P. Saha, Applications of genetic algorithms in QSAR/QSPR modeling, in *Applications of Metaheuristics in Process Engineering*, eds. by J. Valadi, P. Siarry (Springer, 2014), pp. 315–324

4. Y. LeCun, Y. Bengio, G. Hinton, Deep learning. Nature **521**(7553), 436–444 (2015). https://doi.org/10.1038/nature14539

5. J. Lee, L. Xiao, S.S. Schoenholz, Y. Bahri, R. Novak, J. Sohl-Dickstein, J. Pennington, wide neural networks of any depth evolve as linear models under gradient descent, in *33rd Conference on Neural Information Proceeding System* (NeurIPS 2019), Vancouver, Canada

6. R. Gómez-Bombarelli, J.N. Wei, D. Duvenaud, J.M. Hernández-Lobato, B. Sánchez-Lengeling, D. Sheberla, J. Aguilera-Iparraguirre, T.D. Hirzel, R.P. Adams, A. Aspuru-Guzik, Automatic chemical design using a data-driven continuous representation of molecules. ACS Cent. Sci. **4**, 268–276 (2018). https://doi.org/10.1021/acscentsci.7b00572

7. D. Duvenaud, D. Maclaurin, J. Aguilera-Iparraguirre, R. Gómez-Bombarelli, T. Hirzel, A. Aspuru-Guzik, R.P. Adams, Convolutional Networks on Graphs for Learning Molecular Fingerprints. arXiv:1509.09292v2 [cs.LG], 3 Nov 2015

8. T. Xie, J.C. Grossman, Hierarchical visualization of materials space with graph convolutional neural networks. J. Chem. Phys. **149**, 174111 (2018). https://doi.org/10.1063/1.5047803

9. A. Krizhevsky, I. Sutskever, G.E. Hinton, ImageNet classification with deep convolutional neural networks. Adv. NIPS **25**, 1097–1105 (2012)

10. T. Xie, J.C. Grossman, Crystal graph convolutional neural networks for an accurate and interpretable prediction of material properties. Phys. Rev. Lett. **120**, 145301 (2018). https://doi.org/10.1103/PhysRevLett.120.145301

11. A.V. Fedorov, I.V. Shamanaev, Crystal structure representation for neural networks using topological approach. Mol. Inf. **36**, 1600162 (2017). https://doi.org/10.1002/minf.201600162

12. C. Chen, W. Ye, Y. Zuo, C. Zheng, S.P. Ong, Graph networks as a universal machine learning framework for molecules and crystals. Chem. Mater. **31**(9), 3564–3572 (2019). https://doi.org/10.1021/acs.chemmater.9b01294

13. D.B. Kell, S. Samanta, N. Swainston, Deep learning and generative methods in cheminformatics and chemical biology: navigating small molecule space intelligently. Biochem. J. **477**, 4559–4580 (2020). https://doi.org/10.1042/BCJ20200781

14. A. Kadurin, S. Nikolenko, K. Khrabrov, A. Aliper, A. Zhavoronkov, DruGAN: an advanced generative adversarial autoencoder model for de novo generation of new molecules with desired molecular properties in silico. Mol. Pharm. **14**, 3098–3104 (2017)

15. O. Méndez-Lucio, B. Baillif, D.-A. Clevert, D. Rouquié, J. Wichard, De novo generation of hit-like molecules from gene expression signatures using artificial intelligence. Nature Commun. **11**, 1–10 (2020). https://doi.org/10.1038/s41467-019-13807-w

16. R.S. Sutton, Introduction: the challenge of reinforcement learning, in *Reinforcement Learning* (Springer, Boston, MA, 1992), pp. 1–3. https://doi.org/10.1007/978-1-4615-3618-5_1

17. L.P. Kaelbling, M.L. Littman, A.W. Moore, Reinforcement learning: a survey. J. Artificial Intell. Res. **4**(1), 237–285 (1996)

18. S.K. Gottipati, B. Sattarov, S. Niu, Y. Pathak, H. Wei, S. Liu, K. M.J. Thomas, S. Blackburn, C.W. Coley, J. Tang, S. Chandar, Y. Bengio, Learning to navigate the synthetically accessible chemical space using reinforcement learning, in *International Conference on Machine Learning* (PMLR), pp. 3668–3679 (2020)

19. B. Sanchez-Lengeling, C. Outeiral, G.L. Guimaraes, A. Aspuru-Guzik, Optimizing distributions over molecular space, in *An Objective-Reinforced Generative Adversarial Network for Inverse-Design Chemistry (ORGANIC)*. ChemRxiv:5309668.v3

20. M. Popova, O. Isayev, A. Tropsha, *Deep Reinforcement Learning for De-Novo Drug Design*. arXiv:abs/1711.10907 (2017)

21. E. Putin, A. Asadulaev, Y. Ivanenkov, V. Aladinskiy, B. Sanchez-Lengeling, A. Aspuru-Guzik, A. Zhavoronkov, Reinforced adversarial neural computer for de novo molecular design. J. Chem. Inf. Model. **58**, 1194–1204 (2018). https://doi.org/10.1021/acs.jcim.7b00690

22. Y. Khemchandani, S. O'Hagan, S. Samanta, N. Swainston, T.J. Roberts, D. Bollegala, D.B. Kell, DeepGraphMolGen, a multi-objective, computational strategy for generating molecules with desirable properties: a graph convolution and reinforcement learning approach. J. Cheminform. **12**, 53 (2020). https://doi.org/10.1186/s13321-020-00454-3

23. W. Jin, R. Barzilay, T. Jaakkola, Multi-objective molecule generation using interpretable substructures, in *International Conference on Machine Learning* (PMLR), pp. 4849–4859 (2020)
24. A. Zhavoronkov, Y.A. Ivanenkov, A. Aliper, M.S. Veselov, V.A. Aladinskiy, A.V. Aladinskaya, V.A. Terentiev, D.A. Polykovskiy, M.D. Kuznetsov, A. Asadulaev, Y. Volkov, A. Zholus, R.R. Shayakhmetov, A. Zhebrak, L.I. Minaeva, B.A. Zagribelnyy, L.H. Lee, R. Soll, D. Madge, L. Xing, T. Guo, A. Aspuru-Guzik, Deep learning enables rapid identification of potent DDR1 kinase inhibitors. Nature Biotech. **37**, 1038–1040 (2019). https://doi.org/10.1038/s41587-019-0224-x
25. P. Das, T. Sercu, K. Wadhawan, I. Padhi, S. Gehrmann, F. Cipcigan, V. Chenthamarakshan, H. Strobelt, C. dos Santos, P.-Y. Chen, Y.Y. Yang, J.P.K. Tan, J. Hedrick, J. Crain, A. Mojsilovic, Accelerated antimicrobial discovery via deep generative models and molecular dynamics simulations. Nat. Biomed. Eng. **5**, 613–623 (2021)
26. J. Arús-Pous, T. Blaschke, S. Ulander, J.-L. Reymond, H. Chen, O. Engkvist, Exploring the GDB-13 chemical space using deep generative models. J. Cheminform. **11**, 20 (2019). https://doi.org/10.1186/s13321-019-0341-z
27. A. Vaswani, N. Shazeer, N. Parmar, J. Uszkoreit, L. Jones, A.N. Gomez, Ł. Kaiser, I. Polosukhin, Attention is all you need, in *Proceedings of 31st International Conference on Neural Information Proceedings System* (NIPS, Long Beach, CA, USA, 2017), 6000–6010. arXiv: 1706.03762v5 [cs.CL]
28. T.M. Nguyen, T. Nguyen, N. Ho, A.L. Bertozzi, R.G. Baraniuk, S.J. Osher, A primal-dual framework for transformers and neural networks. ICLR 2023
29. V. Bagal, R. Aggarwal, P.K. Vinod, U.D. Priyakumar, MolGPT: molecular generation using a transformer-decoder model. J. Chem. Inf. Model. **62**, 2064–2076 (2022). https://doi.org/10.1021/acs.jcim.1c00600
30. D. Repecka, V. Jauniskis, L. Karpus, E. Rembeza, J. Zrimec, S. Poviloniene, I. Rokaitis, A. Laurynenas, W. Abuajwa, O. Savolainen, R. Meskys, M.K.M. Engqvist, A. Zelezniak, Expanding functional protein sequence spaces using generative adversarial networks. Nature Mach. Intell. **3**, 324–333 (2021). https://doi.org/10.1038/s42256-021-00310-5
31. D. Grechishnikova, Transformer neural network for protein-specific *de novo* drug generation as a machine translation problem. Sci. Rep. **11**, 321 (2021). https://doi.org/10.1038/s41598-020-79682-4
32. N. Ferruz, S. Schmidt, B. Höcker, ProtGPT2 is a deep unsupervised language model for protein design. Nat. Comm. **13**, 4348 (2022). https://doi.org/10.1038/s41467-022-32007-7
33. C. Xu, Y. Wang, A.B. Farimani, TransPolymer: a transformer-based language model for polymer property predictions. NPJ Comput. Mater. **9**, 64 (2023). https://doi.org/10.1038/s41524-023-01016-5
34. J. Devlin, M.-W. Chang, K. Lee, K. Toutanova, BERT: pre-training of deep bidirectional transformers for language understanding, in *Proceedings of 2019 Conference on North American Chap. Association Computer Linguistics: Human Language Technologies*, 1, 4171–4186, Minneapolis, Minnesota. Association for Computational Linguistics. https://doi.org/10.48550/arXiv.1810.04805
35. S. Wang, Y. Guo, Y. Wang, H. Sun, J. Huang, SMILES-BERT: large scale unsupervised pre-training for molecular property prediction, in BCB'19: *Proceedings of 10th ACM International Conference on Bioinformatics, Computation Bio Health Information*, 429–436 (2019). https://doi.org/10.1145/3307339.3342186
36. W. Ahmad, E. Simon, S. Chithrananda, G. Grand, B. Ramsundar, ChemBERTa-2: Towards Chemical Foundation Models. arXiv:2209.01712v1 [cs.LG], 5 Sept 2022
37. C. Kuenneth, R. Ramprasad, PolyBERT: a chemical language model to enable fully machine-driven ultrafast polymer informatics. Nat. Commun. **14**, 4099 (2023). https://doi.org/10.1038/s41467-023-39868-6
38. S. Huang, J.M. Cole, BatteryBERT: a pretrained language model for battery database enhancement. J. Chem. Inf. Model. **62**, 6365–6377 (2022). https://doi.org/10.1021/acs.jcim.2c00035

39. J. Lee, W. Yoon, S. Kim, D. Kim, S. Kim, C. So, J. Kang, BioBERT: a pre-trained biomedical language representation model for biomedical text mining. Bioinformatics **36**(4), 1234–1240 (2020). https://doi.org/10.1093/bioinformatics/btz682

40. Y. Ji, Z. Zhou, H. Liu, R.V. Davuluri, DNABERT: pre-trained bidirectional encoder representations from transformers model for DNA-language in genome. Bioinformatics **37**(15), 2112–2120 (2021). https://doi.org/10.1093/bioinformatics/btab083

41. N. Brandes, D. Ofer, Y. Peleg, N. Rappoport, M. Linial, ProteinBERT: a universal deep-learning model of protein sequence and function. Bioinformatics **38**(8), 2102–2110 (2022). https://doi.org/10.1093/bioinformatics/btac020

42. M. Belkin, D. Hsu, S. Ma, S. Mandal, Reconciling modern machine-learning practice and the bias-variance trade-off. Proc. Nat. Acad. Sci. **116**(32), 15849–15854 (2019). https://doi.org/10.1073/pnas.1903070116

43. M. Belkin, Fit without fear: remarkable mathematical phenomena of deep learning through the prism of interpolation. arXiv:2105.14368v1 (2021)

44. L. Weng, *Are Deep Neural Networks Dramatically Overfitted?* (Lil'Log, 2019). https://lilianweng.github.io/posts/2019-03-14-overfit/

45. J.W. Rocks, P. Mehta, Memorizing without overfitting: bias, variance, and interpolation in overparameterized models. Phys. Rev. Res. **4**, 013201 (2022). https://doi.org/10.1103/PhysReVResearch.4.013201

46. A. Jacot, F. Gabriel, C. Hongler, Neural tangent kernel: convergence and generalization in neural networks, in *Advanced Neural Information Processing System*, 2018, pp. 8580–8589

47. G. Yang, Scaling limits of wide neural networks with weight sharing: Gaussian process behavior, gradient independence, and neural tangent kernel derivation. arXiv:1902.04760 (2019)

48. L. Weng, *Some Math Behind Neural Tangent Kernel* (Lil'Log, 2022). https://lilianweng.github.io/posts/2022-09-08-ntk/

49. J. Lee, Y. Bahri, R. Novak, S.S. Schoenholz, J. Pennington, J. Sohl-Dickstein, Deep neural networks as Gaussian processes. arXiv:1711.00165v3 (2018)

50. Z. Liu, Y. Wang, S. Vaidya, F. Ruehle, J. Halverson, M. Soljačić, T.Y. Hou, M. Tegmark, *KAN: Kolmogorov–Arnold Networks*. arXiv:2404.19756v2 [cs.LG], 2 May 2024

51. A.N. Kolmogorov, On the representation of continuous functions of several variables as superpositions of continuous functions of a smaller number of variables. Dokl. Akad. Nauk **108**(2), (1956)

52. A. Anandkumar, R. Ge, D. Hsu, S.M. Kakade, M. Telgarsky, Tensor decompositions for learning latent variable models. arXiv:1210.7559 (2014)

Chapter 7
Discovery and Creativity

> *Every discovery contains an irrational element or a creative*
> *intuition.*
> —*Karl Popper* [1]
> *Creativity is just connecting things.*
> *Steve Jobs* [2]

Unlike the material in the rest of this book, the contents of this chapter will necessarily be more subjective, qualitative, speculative and perhaps even controversial, reflecting my own personal experiences, possible biases and judgements, and will stray somewhat from the realm of chemistry and materials science. They are included here to provide food for thought and reflection. Before we discuss creativity in the context of AI, let us consider what we understand by human creativity. There will perhaps never be perfect agreement on this point, and therefore it would be naïve to expect such agreement on the question of whether and when AI can be regarded as being creative. As a part-time artist, I have long argued against the trend, prevalent in the art world throughout much of the last century, of equating mere novelty and shock value with artistic creativity.

There are only a finite number of words in the English language, or in any language; so nothing is "new" or really creative in the sense of creating new words; it is only the rearrangement of the finite corpus of words that creates novelty. But if novelty were the sole measure of creativity, then a group of monkeys randomly typing out text could be considered creative! Obviously, this is too low a bar. We require the output text to make sense—to be structured into words, sentences, coherent thoughts, ideas or concepts. But even this would be too high a bar for some. There is a genre of literature known as "stream of consciousness writing" that dispenses with some of these requirements, simply providing a running monologue into what transpires in the minds of the characters. Similar trends exist in the art and music scenes, where art is sometimes equated with self-expression and nothing more.

We can disregard these fashionable trends in the art world when it comes to generative chemistry. The equivalent to "making sense" is that the molecules generated by AI should make chemical sense—in the realm of generative chemistry this is called *validity* (whether the generated molecules correspond to valid chemical structures), and it is quite distinct from the concept of *novelty* (whether or not the valid generated

N. Sukumar, *Navigating Molecular Networks*,
SpringerBriefs in Materials, https://doi.org/10.1007/978-3-031-76290-1_7

molecules are outside the training set of the generative model). But mere validity and novelty are not enough for most practical applications. Any novel, valid molecule need not be an effective drug or battery material. As we have seen, generative models in chemistry are often evaluated on the basis of validity, novelty, uniqueness (the fraction of valid generated molecules that are unique), internal diversity (assessed by a similarity metric) and, for drug discovery applications, Quantitative Estimate of Drug-likeness (QED)—often termed "chemical beauty."

I have deliberately used the expression "coherent thoughts, ideas or concepts" above. Many people, both within and outside the AI community, often dispute whether large language models (LLMs) can even have "thoughts, ideas or concepts", and whether they "understand" thoughts, ideas and concepts. It has been argued that these models are merely "stochastic parrots," essentially repeating probable answers without any comprehension [3]. Certainly this critique holds some merit, as LLMs operate by assigning probabilities to words and tokens. But let us delve deeper into what we really mean by this word: "understand."

Consider what it means to understand fields like chemistry, physics, biology, or society. Understanding chemistry involves grasping concepts such as the periodic table, electronegativity, and molecular interactions. Understanding physics entails knowledge of forces, equations of motion, and principles like spacetime. Understanding biology includes comprehending natural selection, bodily functions, and protein interactions. Understanding society means having insights into human relationships and societal dynamics. Such understanding consists of having an internal representation of the world, sufficient to be able to identify connections, extract patterns, and make useful predictions.

Does AI possess a similar level of understanding about the vast datasets it has been trained on? Undoubtedly, yes. DL models do have an internal "latent" representation of the data, and are able to identify connections therein and extract patterns from the data. This "understanding" enables these models to be useful and predictive. However, this does not mean that AI "understands" physics, chemistry or biology in the sense that we do. AI understanding is not analogous to human understanding. Nevertheless, AI is able to make functional predictions within its domain of applicability. Forcing AI models to mimic human learning processes is not particularly effective. Occasionally, these models align with human-conceived concepts, making them interpretable in human terms, and this is the field of interpretable AI. Often this interpretability comes at the cost of some (acceptable) sacrifice in predictivity. Yet, more often than not, deep learning models remain opaque to us.

As we have discussed on Sect. 5.3, any model is valid for making predictions only within its domain of applicability, and can be expected to fail spectacularly when misapplied far outside this range. Similar uncertainties apply as well to human models and human understanding. Both AI and human models can be biased or fail to predict outcomes accurately as new data emerges. Much of our current understanding of the physical universe and biology is continually evolving and being refined through scientific inquiry. Similarly, reinforcement learning (RL) models operate by learning from their mistakes, somewhat akin to how scientific knowledge advances.

The unhelpful term "hallucination" is widely used to describe LLMs "confidently" making spectacularly wrong predictions. First of all, it should be pointed out that the use of the term "confident" to describe LLMs is a human projection onto an algorithm. In all likelihood, the LLM did not provide confidence estimates (as, for example, in GPR) and was not asked to do so. It was (implicitly) prompted to provide the most probable completion based on its training set. Most failures of AI models are due to either bias in the training data or misapplication outside their domain of applicability. LLMs contain a "temperature" parameter that determines the probability of producing a less likely answer. At zero temperature, the LLM will always generate only the most probable token at any position, leading to a repetitive, deterministic (and thus "uncreative") answer. As we dial up the temperature parameter, the probability of selecting the less likely tokens at any position increases, thereby increasing the tendency to produce novel, "creative" output, or for a chemical LLM, to explore novel regions of chemical space, but also increasing the probability of "hallucinations". So is AI creativity just a matter of dialing up the temperature parameter? I leave that question for the reader to ponder over.

Certainly there are differences between artistic creativity and scientific creativity, but in my opinion, not as much as is commonly believed. Scientific creativity is constrained by the laws of physics and the rules of mathematics. Artistic creativity is not unconstrained either; it is constrained by the laws of language (in the case of poetry and literature), by the limitations of the medium (in the case of art and music), and even more so, by the cultural environment. Of course, all these constraints can be broken, and truly creative artists do so, to create new genres of music, art and poetry. Similarly, creative science comes with questioning the fundamental axioms, extending the known boundaries, giving rise to what Kuhn [4] termed new paradigms. This can conflict with the notion of validity mentioned above. Discovery in chemistry and materials science often involves, whether by design or serendipity, generating new classes of molecules, structures or materials, going well beyond the set of training molecules and therefore well outside the domain of applicability of generative models. For instance, several years ago, when we were developing ML models to design novel polymer dielectrics, we came up with a new class of tin-based polymers [5] because we had failed to exclude inorganics from our virtual high-throughput screen. Therefore I consider some degree of "hallucination" an inevitable by-product of creativity. Where would we be without the hallucinations of van Gogh, Gauguin, Edvard Munch, William Blake, Hildegard of Bingen, or the visions of Ramanujan?

So I consider the debate over whether LLMs "understand" to be rather unproductive. LLMs construct a functional model of their domain, albeit in a manner distinct from human cognition. It is a linguistic choice whether or not we want to consider this a form of understanding. Dogs, for instance, perceive the world very differently from humans but still navigate it effectively. Dogs don't understand Shakespeare or quantum physics, but they do understand human emotions and their own surroundings well enough to serve their purposes and enable them to survive and thrive in this world. But unlike LLMs, both dogs and humans are biological beings with brains that share a common evolutionary origin. So whether we choose to restrict the term "understanding" to biological organisms or extend it to include AI is, in my opinion, largely a matter of terminology rather than a reflection of fundamental functional capabilities.

This is, of course, not to say that there are no fundamental functional differences between human/biological understanding and machine intelligence. A human being is not a disembodied brain. A disembodied brain cannot learn from its environment. In order for AI to become more "human-like" in its capabilities, it has to able to interact with the real world in real time. This is accomplished by retrieval-augmented generation (RAG) [6], a technique wherein generative AI models are augmented, as and when required, with new data retrieved from external sources, in order to improve the accuracy and reliability of the models. In the realm of generative chemistry/materials science, active learning models [7] integrated with robotic tools are being used to perform syntheses and experimental characterizations of new molecules [8], or interactively query the user to label new molecules with the appropriate activities/properties, so as to use this enhanced dataset to train improved models. Such active learning strategies, with iterative loops of model training, validation and characterization, followed by improvement of the ML models, will be essential for extending the original domain of applicability and probing hitherto unexplored regions of chemical space.

References

1. *Top 140 Karl Popper Quotes.* Quotefancy (2024). https://quotefancy.com/karl-popper-quotes
2. *21 Inspiring Creativity Quotes That'll Get Your Ideas Flowing.* WeWork Companies LLC (2024). https://www.wework.com/ideas/professional-development/creativity-culture/creativity-quotes
3. N. Sukumar, *Does AI "Understand"?* LinkedIn Corporation, October 5 (2023). https://www.linkedin.com/pulse/does-ai-understand-n-sukumar/
4. T. Kuhn, *Structure of Scientific Revolutions* (University of Chicago Press, Chicago, 1962)
5. G. Pilania, C. Wang, K. Wu, N. Sukumar, C.M. Breneman, G.A. Sotzing, R. Ramprasad, New group IV chemical motifs for improved dielectric permittivity of polyethylene. J. Chem. Inf. Model. **53**(4), 879–886 (2013). https://doi.org/10.1021/ci400033h
6. P. Lewis, E. Perez, A. Piktus, F. Petroni, V. Karpukhin, N. Goyal, H. Küttler, M. Lewis, W. Yih, T. Rocktäschel, S. Riedel, D. Kiela, Retrieval-augmented generation for knowledge-intensive NLP tasks. NIPS'20: Proc. 34th Int. Conf. Neural Inf. Proc. Sys. **793**, 9459–9474 (2020)
7. A. Merchant, S. Batzner, S.S. Schoenholz, M. Aykol, G. Cheon, E.D. Cubuk, Scaling deep learning for materials discovery. Nature **624**, 80–85 (2023). https://doi.org/10.1038/s41586-023-06735-9
8. N.J. Szymanski, B. Rendy, Y. Fei, R.E. Kumar, T. He, D. Milsted, M.J. McDermott, M. Gallant, E.D. Cubuk, A. Merchant, H. Kim, A. Jain, C.J. Bartel, K. Persson, Y. Zeng, G. Ceder, An autonomous laboratory for the accelerated synthesis of novel materials. Nature **624**, 86–91 (2023). https://doi.org/10.1038/s41586-023-06734-w

Glossary of Terms

Activation function A function that transforms the weighted sum of the inputs a neuron receives into the output of the neuron Eq. 6.1

Active learning An ML strategy where the learning algorithm can interactively query the user to label new molecules with the desired outputs, and use this enhanced dataset to train improved models

Activity Cliff A discontinuity in the structure–activity landscape, corresponding to a large change in biological activity for a relatively small change in molecular structure

Activity cliff generators Molecules with a high probability of forming activity cliffs with other molecules tested in the same biological assay

Adjacency matrix A An $n \times n$ matrix representation of a graph, with each row and each column representing a different node, whose elements are 1 (for an unweighted graph) if the corresponding nodes share an edge, and 0 otherwise Eq.1.1 and 1.2 Table 1.1a

Algebraic connectivity a(G) The second-smallest eigenvalue (counting multiple eigenvalues separately) of the Laplacian matrix of a graph Table 3.1

Artificial Neural Network (ANN) A machine learning paradigm patterned on the neural connections in the brain, consisting of layers of nodes (neurons) connected to each other with associated weights that can be modified during training, and where each neuron computes some function of the weighted sum of its inputs Fig. 6.2

Attention A mechanism that captures the correlations between different tokens in a string, regardless of their relative position, to decide which of them are important for predicting the next token Eq. 6.7

Auto Encoder (AE) A feed forward neural network, consisting of an encoder with a bottleneck architecture that is trained to map (encode) the input into a low-dimensional latent vector, and a decoder with the opposite architecture that reconstructs the input from the latent vector Fig. 6.4

Average clustering coefficient $C(G)$ The clustering coefficient of nodes averaged over all the nodes in the graph Eq. 3.8

© The Editor(s) (if applicable) and The Author(s), under exclusive license 107
to Springer Nature Switzerland AG 2024
N. Sukumar, *Navigating Molecular Networks*,
SpringerBriefs in Materials, https://doi.org/10.1007/978-3-031-76290-1

Average degree The average of the degrees over the nodes of the network Eq. 3.2

Average path length l(G) The shortest path connecting pairs of nodes, averaged over all pairs of vertices np in a graph Eq. 3.6

Back propagation The common process of learning in a supervised ANN, where the output value of the output layer is compared to the target value, and the discrepancy between the two (error or loss) is propagated back to appropriately modify the weights and biases in previous layers

Basis A set of linearly independent vectors spanning a vector space

Betweenness centrality C_B A measure of the frequency of occurrence of a given node on the shortest path between all pair of nodes in a graph Eq. 3.59 Fig. 3.6

Bipartite graph A graph whose nodes belong to two distinct classes, where an edge may only connect a node of one class to a node of the other class

Brody parameter β A parameter distinguishing different statistical ensembles (β = 0: Poisson, β = 1: GOE, β = 2: GUE, β = 4: GSE)

Chemical space An abstract multi-dimensional space, whose coordinate axes are molecular descriptors, with different regions of the space populated by molecules sharing some chemical similarity

Chemical space similarity network (CSN) A network whose nodes represent distinct molecules, and whose edges connect pairs of similar molecules Fig. 1.5b

Classification An ML task where the available data includes only binary or categorical labels, and we wish to predict the labels for the test set

Closeness centrality C_C The reciprocal of the sum of the lengths of the shortest paths to every other node, a measure that describes the closeness of a node to all other nodes in a graph Eq. 3.58

Clustering coefficient C_{iG} The number of triangles which pass through a node divided by the total number of possible triangles through the node Eq.3.7

Community structure The existence of densely connected groups of nodes or communities in a network, with a large number of edges between the nodes within a community, but much sparser connections between different communities

Complete graph A graph where every pair of nodes is directly connected by an edge

Component A connected subgraph that is not part of any larger connected subgraph

Connected graph A graph where there is a path between every pair of nodes

Convolutional Neural Network (CNN) A type of neural network whose architecture was inspired by the visual cortex of the mammalian brain, comprised of one or more convolutional layers, followed by one or more fully connected layers Fig. 6.5

Cost function See loss function

Curse of dimensionality The loss of generalizability resulting from overfitting when the number of training examples is vastly exceeded by the number of descriptors in the model

Deep Neural Network (DNN) A neural network composed of many hidden layers, where each layer transforms the representation from the previous layer into a representation at a higher, more abstract level

Degree The number of connections a given node has with other nodes in a graph Eq. 1.3

Degree assortativity The correlation coefficient between the degrees of connected nodes in a graph

Degree centrality A measure of the number of links a node has in a graph Eq. 3.57

Degree centralization The average deviation of node degrees from the maximum degree, normalized to its maximum value for a star graph Fig. 3.5

Degree distribution P(k) The probability that a randomly selected node in the network has degree k

Diameter The longest path length in a graph

Directed graph A graph whose edges are associated with a direction, and represented by an adjacency matrix that is not symmetric about the main diagonal Fig. 1.1c

Discriminator The component of a GAN that attempts to distinguish the input data from those generated by the generator

Edge connectivity e(G) The minimum number of edges whose removal would result in loss of connectivity of the graph Table 3.1

Edge density $\rho(G)$ The ratio of the number of edges to the maximum number of possible edges in a graph Eq. 3.5 Fig.3.1

Eigenvector centrality A graph measure based on spectral analysis of graphs, that ranks nodes in order of the eigenvalues of the associated matrices Eq. 3.63

Empty graph A graph with no edges

Encoding A transformation of the original data to a new vector space of lower dimensionality, while preserving its essential features

Erdős-Rényi (ER) random network A network with edges that are chosen randomly with a given probability

Feature selection Selecting a subset of independent features for machine learning

Feed-forward neural network A neural network where each neuron receives inputs only from neurons in the previous layer, and feeds its output forward to neurons in the next layer

Fiedler eigenvalue See algebraic connectivity

Forman-Ricci curvature A discretization of the Ricci curvature tensor, that quantifies the flow along an edge in a network, weighted by the adjacent edges Fig. 3. 7 Eqs. 3.65 and 3.66

Gaussian orthogonal ensemble (GOE) A Gaussian ensemble for time-reversal invariant systems with rotational symmetry, with real, symmetric Hamiltonian matrix, characterized by $\beta = 1$

Gaussian Process Regression (GPR) A probabilistic regression method based on the assumption that the function to be predicted is drawn from a Gaussian process, allowing for uncertainty quantification along with the predictions Fig. 2.3

Gaussian unitary ensemble (GUE) A Gaussian ensemble for systems that violate time-reversal invariance, with Hermitian Hamiltonian matrix, characterized by $\beta = 2$

Gaussian symplectic ensemble (GSE) A Gaussian ensemble for time-reversal invariant systems with half-integer spin and broken rotational symmetry, characterized by $\beta = 4$

Generative Adversarial Network (GAN) An AI architecture inspired by game theory, that consists of two DNNs—a Generator and a Discriminator, that compete against each other in a machine learning game Fig. 6.6

Generator The component of a GAN that generates new data matching the statistics of the input data, and tries to fool the discriminator into believing that the generated data are actually derived from the input dataset

Genetic Algorithm (GA) A stochastic optimization method, belonging to the class of evolutionary computation, that mimics how a population of chromosomes evolves through the processes of cross-over and random mutation in natural selection

Giant component A component that is much larger than all others in the graph

Graph A collection of nodes or vertices, and pairwise relationships represented by edges connecting the nodes in pairs

Graph Convolutional Neural Network (GCNN) A neural network built on the molecular graph of a molecule or crystal, with node, edge and global attributes

Graph spectrum The set of eigenvalues of the adjacency matrix or Laplacian matrix of a graph

Hierarchical network A class of networks showing coexistence of modularity, local clustering and scale-free topology, where communication between different clusters is mediated by a few hubs

Hubs High-degree nodes in a network

Homophily The tendence of similar nodes in a network to cluster together

Incidence matrix M An $n \times m$ matrix representation of a graph, with rows indexed by the nodes and columns by the edges, such that $M_{ij} = 1$ if node i is connected to edge j, and $= 0$ otherwise Table 1.1.1d Eq. 1.6

Inner product a generalization of a dot product of two vectors Eq. 2.11

Kernel function A function that defines a similarity measure between the original vectors Eq. 2.28

Kernel Partial Least Squares (KPLS) A nonlinear generalization of PLS Eq. 2.53

Kernel trick A transformation that maps a complicated non-linear similarity relationship between molecules in the original descriptor space into a dot product relationship in some higher dimensional space

Kohonen map An unsupervised artificial neural network that employs a process of competitive learning between nodes

Kolmogorov-Arnold Network (KAN) An ANN with learnable activation functions on its edges, where each layer is a matrix of univariate functions Eq. 6.25–6.29 Fig. 6.10

Kriging See Gaussian Process Regression

Laplacian matrix L An $n \times n$ matrix whose diagonal elements are the node degrees and off-diagonal elements are the negative of the corresponding adjacency matrix elements Table 1.1c Eq. 1.5

Ligand-based drug design A computational drug design strategy undertaken without knowledge of the target structure, by exploiting similarities between molecules known to cause the desired biological activity

Linear transformation A mapping from one vector space to another that respects the underlying linear structure of each vector space Eqs. 2.9 and 2.10

Logistic function See sigmoid function Fig.6.3a Eq. 6.3

Loss function A function that computes the discrepancy between the output value of a neuron and the target value

Metric tensor A structure defined on a manifold that allows for the definition of distances and angles

Model Applicability Domain The (chemical) space on which the training set of the model has been developed, and for which it is applicable to make predictions for new molecules

Modularity Q A quantitative measure of community structures in a network Eq. 3.9

Molecular descriptor Numerical or categorical representations of molecules

Multi-layer perceptron (MLP) A feed-forward ANN consisting of fully connected neurons with a nonlinear activation function

Nearest-neighbor spacing distribution (NNSD) The distribution of spacings between neighboring eigenvalues

Next-nearest-neighbor spacing distribution (nNNSD) The distribution of spacings between next nearest neighbor eigenvalues

Neural tangent kernel (NTK) A kernel describing the evolution of a DNN during training by gradient descent Eq. 6.16

Oriented incidence matrix N An $n \times m$ matrix representation of a graph, with rows indexed by the nodes and columns by the edges, such that $N_{ij} = 1$ if node i is the 'head' of edge j, $= -1$ if node i is the 'tail' of edge j, and $= 0$ otherwise Table 1.1f Eq. 1.14

Overfitting The tendency of a model to "memorize" the training data, instead of learning to generalize, resulting in perfect or excellent performance on the training set, but poor performance on hitherto unseen data

Partial Least Squares (PLS) regression A multivariate regression method combining PCA with a regression step in an iterative procedure Eq. 2.20

Path length A measure the defines the shortest path between a pair of nodes in a graph

Poisson degree distribution A degree distribution following Poisson statistics, falling off asymptotically as: $P(k) \sim e^{-k}$ Eq. 3.10

Power law degree distribution A degree distribution that decays as a power law with the degree k: $P(k) \sim k^{-\gamma}$ Eq. 3.12 Fig.3.4

Preferential attachment The tendency of new nodes to attach preferentially to nodes of high degree when added to a network

Principal Component Analysis (PCA) A multivariate data analysis technique, used to reduce the dimensionality of a data set, while retaining its information content. A linear transformation of the original data vectors such that: the first principal component explains the maximum variance in the data set, with each subsequent component describing the maximum part of the remaining variance, subject to the condition that all principal components are orthogonal to each other Fig. 2.1b and Fig. 5.1a Eq. 2.17

QSAR Quantitative Structure Activity Relationship

QSPR Quantitative Structure Property Relationship

Random Matrix theory (RMT) A theory developed by Eugene Wigner to describe statistical aspects of systems with many degrees of freedom, replacing the system Hamiltonian by an ensemble of random matrices, to focus attention on generic properties determined by the underlying symmetries, leading to the emergence of universal laws

Ratio distribution The ratio of consecutive spacings between eigenvalues Eq. 4.23

Rectified linear unit (ReLU) An activation function that functions as a half-wave rectifier Fig. 6.3b Eq. 6.5

Recurrent neural network (RNN) A neural network wherein the output of a neuron can also be one of its own inputs, giving rise to self-loops

Regression An ML task where the training set has numerical properties, and we wish to predict the numerical values of the target properties for the test set

Reinforcement Learning An ML paradigm that eschews the use of brute-force to examine all possible solutions, but instead discovers the actions that yield the maximum reward by interacting with a dynamic environment and learning an optimal policy by trial and error

Ricci curvature tensor A tensor that quantifies the extent of deformation of the space from a locally Euclidean geometry as one moves along geodesics in the space

Riemannian metric tensor A metric tensor on a Riemannian manifold Eq. 2.14

Scaffold Hopping The ability of a model to generalize across diverse structural scaffolds

Scale-free network A network characterized by a degree distribution that decays as a power law with the degree Eq. 3.12

Self-Attention A mechanism that provides context for every position (token) in a string by computing correlations between different tokens in the string and deciding which ones are important for that token Eq. 6.7

Self-Organized Map (SOM) See Kohonen map

Sigmoid function A smooth analytic alternative to the step function Fig. 6.3a Eq. 6.3

Similarity principle The principle that molecules with similar structures exhibit similar properties, while molecules with dissimilar structures exhibit dissimilar properties

Simple graph A graph without any self-loops connecting a node to itself

Small-world property/network One with a high clustering coefficient and a characteristic path length that increases slowly as a function of the number of nodes in the network

Softmax activation function A generalization of the logistic function to multiple dimensions, that converts a vector of real numbers into a probability distribution of possible outcomes Eq. 6.4

Spectral rigidity Δ3 The deviation of the spectral staircase function from the best fitting straight line, a measure of long-range correlation between eigenvalues, or "stiffness" of the eigenvalue spectrum Eq. 4.27

Spectral staircase function η(x) The number of levels in the interval $[-\infty, x]$ on the unfolded scale Eq. 4.26

Structure-activity landscape index (SALI) A quantitative measure, designed by Rajarshi Guha and John Van Drie, characterizing the steepest cliffs or discontinuities in a structure-activity landscape Fig. 5.2

Subgraph A graph formed from a subset of the nodes and edges of G a larger graph

Subgraph centrality A graph measure obtained by summing the number of closed walks of various lengths starting and ending at a node, giving higher weights to the smaller subgraphs Eqs. 3.61 and 3.62

Supervised learning An ML paradigm wherein a model is first trained on a training set with labeled data, such as molecules whose activities have been experimentally measured, and the trained model is then used to predict the activities of other test molecules

Support Vector Machines (SVM) An ML method used for both classification and regression, that relies on maximizing (for classification)/maximizing (for regression) the margin or separation between two auxiliary hyperplanes Fig. 2.2 Eqs. 2.31–2.52

Support Vectors Data points lying on the auxiliary hyperplanes in SVM Eq. 2.32

t-distributed Stochastic Neighbor Embedding (t-SNE) A nonlinear, iterative, stochastic dimensionality reduction algorithm that maps data from a high dimensional to a lower dimensional space, while retaining its local structure Fig. 5. 1b

Transfer learning An ML strategy of pre-training on large datasets for simple properties, followed by fine-tuning with the limited available data on the desired property

Transformer A deep neural network architecture exploiting the self-attention mechanism, and consisting of several multi-head attention layers and fully connected layers Fig. 6.4

Uniform Manifold Approximation and Projection (UMAP) A nonlinear, graph-based dimensionality reduction algorithm that strikes a balance between preserving local versus global structure in the data

Undirected graph A graph whose edges are not associated with a direction, and represented by an adjacency matrix that is symmetric about the main diagonal

Unfolding A transformation of eigenvalues to constant spectral density on average Eq. 4.5

Unipartite graph A graph whose nodes all belong to the same class, where any node may be connected to any other through an edge

Unsupervised learning An ML paradigm with unlabeled data, where we do not have a training set with known activities, and the data are grouped into different classes based solely on the distribution and statistical properties of the dataset

Variational Auto Encoder (VAE) An auto-encoder where the latent space points are replaced by a continuous probability distribution

Vector space A space of vectors with the properties of associativity of vector addition, commutativity of vector addition, identity element of vector addition, inverse elements of vector addition, compatibility of scalar multiplication with field multiplication, identity element of scalar multiplication, distributivity of scalar multiplication with respect to vector addition and field addition, defined on it Eq. 2. 1– 2.8

Vertex connectivity v(G) The minimal number of nodes together with their adjacent edges whose removal would result in loss of connectivity of the graph Table 3.1

Watts-Strogatz network A network formed by starting with nodes on a ring with only local edges, and then randomly rewiring each edge with a given probability

Weighted graph A graph where each edge is associated with a weight, and the matrix elements of the weighted adjacency matrix can take on any real number value associated with the weights

Wide neural network A neural network that has a large number of neurons in each hidden layer

Wigner surmise A form for the nearest-neighbor eigenvalue spacing distribution (NNSD) Eq. 4.8